Channel-Based Multi-Queue SBC Process Algebra For Systems Definition

-- Integration of Systems Structure and Systems Behavior --

Second Edition

William S. Chao

Structure-Behavior Coalescence

Systems Architecture = Systems Structure + Systems Behavior

CONTENTS

PREFACE TO THE SECOND EDITION

In the first edition of this book, we do not treat any conditional expression. Therefore, all the theories of channel-based multi-queue SBC process algebra (C-M-SBC-PA) attempt to ignore all conditional expressions.

After further research, we find that if we embed each condition into a prefix, all the theories of channel-based multi-queue SBC process algebra will be more complete and beautiful.

In the second edition of this book, the reader would enjoy to probe conditional expressions that are so elegantly handled in every aspect of channel-based multi-queue SBC process algebra.

PREFACE TO THE FIRST EDITION

The need for systems definition arises because any real-life system is inherently complicated. It is impossible to comprehend fully the intricate interaction of any system of the real world with its environment, or to define all its components and each of its details. Systems definition or system definition is an artifact created by humans to describe what a system is.

Process algebras are a diverse family of related approaches to the study of concurrent systems. Their tools are algebraic languages for the high-level description of interactions, communications and synchronizations between a collection of independent agents or processes. Process algebras also provide algebraic laws that allow process descriptions to be manipulated and analyzed, and permit formal reasoning about equivalences and observation congruence among processes. Accordingly, process algebra provides a perfect method for systems definition.

Channel-based multi-queue SBC process algebra (C-M-SBC-PA) is one of the six specialized SBC process algebras. In this book, we use C-M-SBC-PA to achieve the robust systems definition of a system. To see is to believe. Therefore, many examples are presented to help the reader fully understand the use of C-M-SBC-PA.

ABOUT THE AUTHOR

Dr. William S. Chao is the CEO & founder of SBC Architecture International®. SBC (Structure-Behavior Coalescence) architecture is a systems architecture which demands the integration of systems structure and systems behavior of a system. SBC architecture applies to hardware architecture, software architecture, enterprise architecture, knowledge architecture and thinking architecture. The core theme of SBC architecture is: "Architecture = Structure + Behavior."

William S. Chao received his bachelor degree (1976) in telecommunication engineering and master degree (1981) in information engineering, both from the National Chiao-Tung University, Taiwan. From 1976 till 1983, he worked as an engineer at Chung-Hwa Telecommunication Company, Taiwan.

William S. Chao received his master degree (1985) in information science and Ph.D. degree (1988) in information science, both from the University of Alabama at Birmingham, USA. From 1988 till 1991, he worked as a computer scientist at GE Research and Development Center, Schenectady, New York, USA.

Dr. William S. Chao has been teaching at National Sun Yat-Sen University, Taiwan since 1992 and now serves as the president of Association of Enterprise Architects, Taiwan Chapter. His research covers: systems architecture, hardware architecture, software architecture, enterprise architecture, knowledge architecture and thinking architecture.

PART I: BASIC CONCEPTS

Chapter 1: Introduction to Systems Definition

All things that amaze us as something independent are essentially parts of a system. We usually call the parts of a system its components. Every system is something the whole. Systems emphasize the holistic vision.

The need for systems definition arises because any real-life system is inherently complicated. It is impossible to comprehend fully the intricate interaction of any system of the real world with its environment, or to define all its components and each of its details. Systems definition or system definition is an artifact created by humans to define what a system is.

In this chapter, we first introduce physical and conceptual systems. A physical system exists in the physical, concrete, or real world. A conceptual system exists in the conceptual, abstract, or virtual world. We then introduce the boundary of a system. A system has a boundary. The system itself is inside the boundary and the environment is outside the boundary. After these introductions, we shall thereafter discuss the systems definition 1.0 as well as the systems definition 2.0.

1-1 Physical and Virtual Systems

In general, systems are divided into two categories: 1) physical systems and 2) virtual systems.

A physical system exists in the physical world [Acko68]. A physical system is also called a concrete or real system. For example, a *telephone* composed of *microphone, earphone and keypad*, shown in Figure 1-1, is a physical, concrete, or real system.

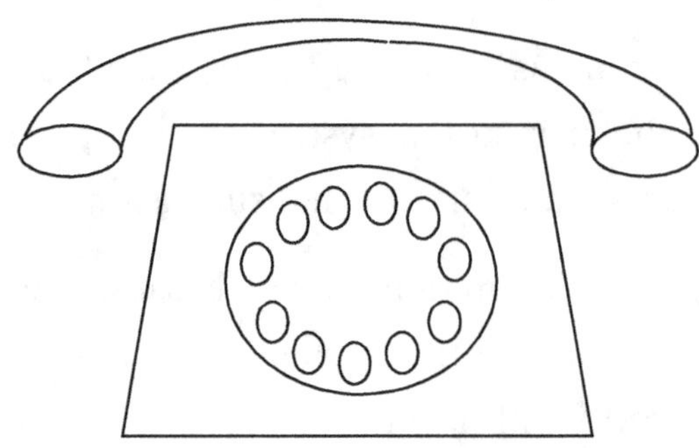

Figure 1-1 A Telephone is a Physical System

As a second example, a *stool* composed of *seat* and *legs*, shown in Figure 1-2, is a physical, concrete, or real system.

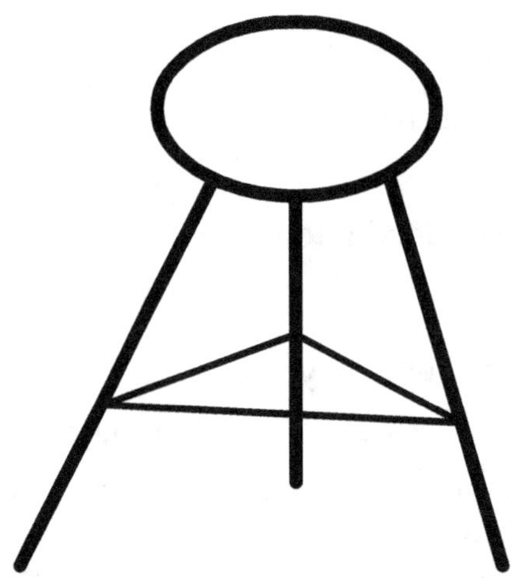

Figure 1-2 A Stool is a Physical System

A virtual system is a system that is composed of non-physical components, i.e., ideas, thoughts, or notions. A virtual system exists in the virtual, abstract, or notional world. For example, the "*Snow White and the Seven Dwarfs*" fairy tale composed of "*Snow White*" and "*Seven Dwarfs*," shown in Figure 1-3, is a virtual, abstract, or notional system.

Figure 1-3 *Snow White and the Seven Dwarfs* is a Virtual System

As a second example, the "*Multi-Tier Personal Data System*" software composed of *MTPDS_GUI*, *Age_Logic*, *Overweight_Logic* and *Personal_Database*, shown in Figure 1-4, is a virtual, abstract, or notional system.

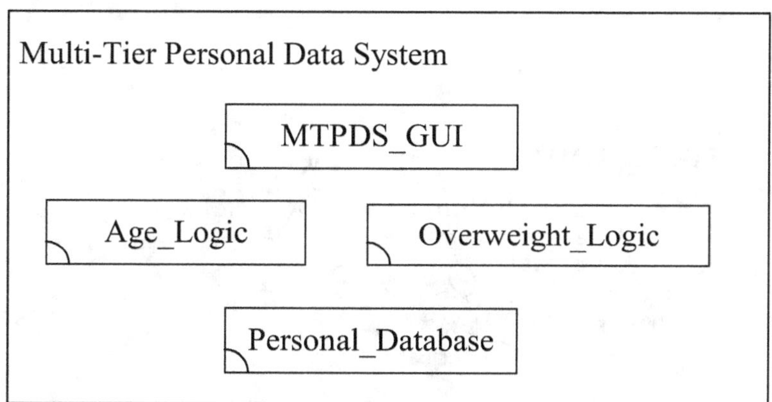

Figure 1-4 *Multi-Tier Personal Data System* is a Virtual System

1-2 Boundary and Environment of a System

We scope a system by describing its boundary as shown in Figure 1-5. All components of the system are inside the boundary while the environment is outside the boundary.

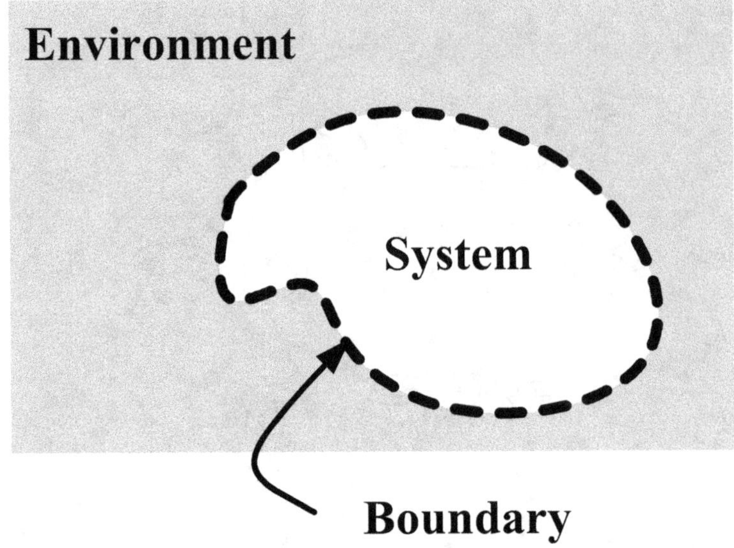

Figure 1-5 Boundary and Environment of a System

The environment is also known as the surroundings. A system may or may not interrelate with the environment. An open system interrelates with the environment through the exchange of matter, energy, data, information, or message as shown in Figure 1-6.

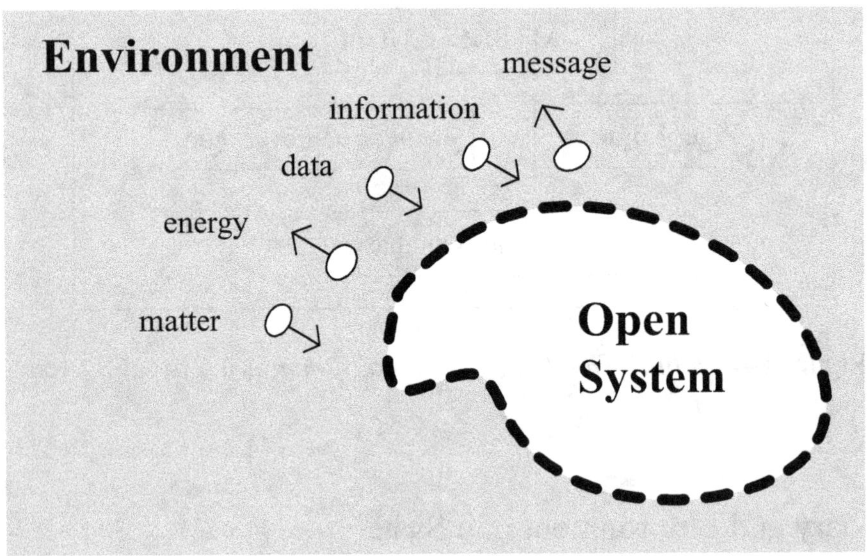

Figure 1-6 Open System Interrelates with the Environment

An isolated system does not interrelate with the environment at all. There is no exchange of matter, energy, data, information, or message between the isolated system and the environment as shown in Figure 1-7.

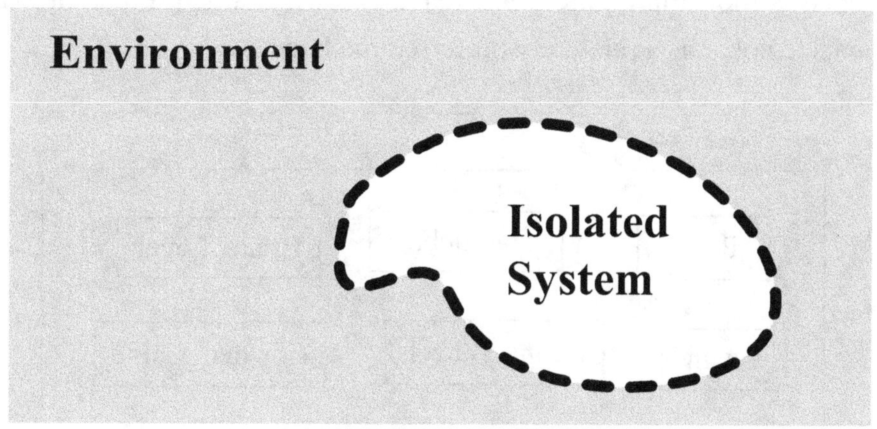

Figure 1-7 Isolated System Does Not Interrelate with the Environment

1-3 Systems Definition 1.0

Systems definition 1.0 defines a system, in Figure 1-8, hopefully to be an integrated whole, embodied in its assembled components, their interrelationships with each other and the environment [Chec99, Frie11, Ghar11, Mead08].

A system, hopefully is an integrated whole, embodied in its assembled components, their interrelationships with each other and the environment.

Figure 1-8 Systems Definition 1.0 Defining a System

Components are sometimes labeled as parts, entities, objects, building blocks and non-aggregated systems [Chao14a, Chao14b, Chao14c]. Interrelated components make a system not only a whole but also hopefully an integrated whole.

A system described by systems definition 1.0 has the following characteristics: 1) hopefully, it is an integrated whole; 2) it is embodied in its assembled components;

3) components are interrelated with each other and the environment; and 4) it uses structural decomposition [Chao14b, Ghar11] rather than functional decomposition [Scho10].

The structural decomposition method is to decompose a system into a number of components, as shown in Figure 1-9. Breaking down a large problem into a number of components to solve, is a relatively preferred method.

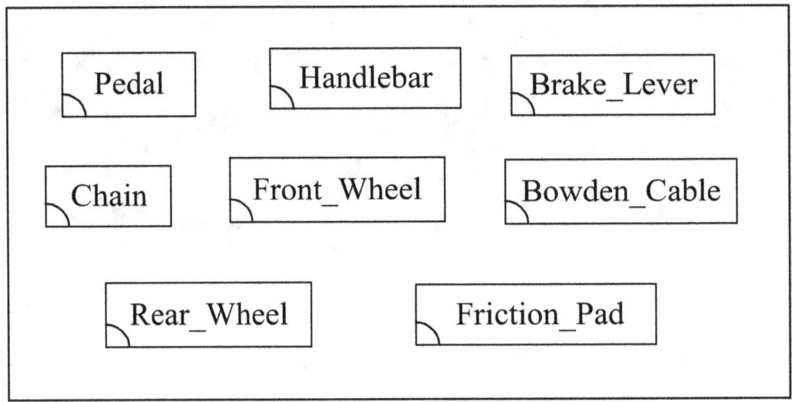

Figure 1-9 Structural Decomposition Method

The functional decomposition method is to decompose a system into a number of functions, as shown in Figure 1-10. Breaking down a large problem into a number of functions to solve, is a relatively non-preferred method.

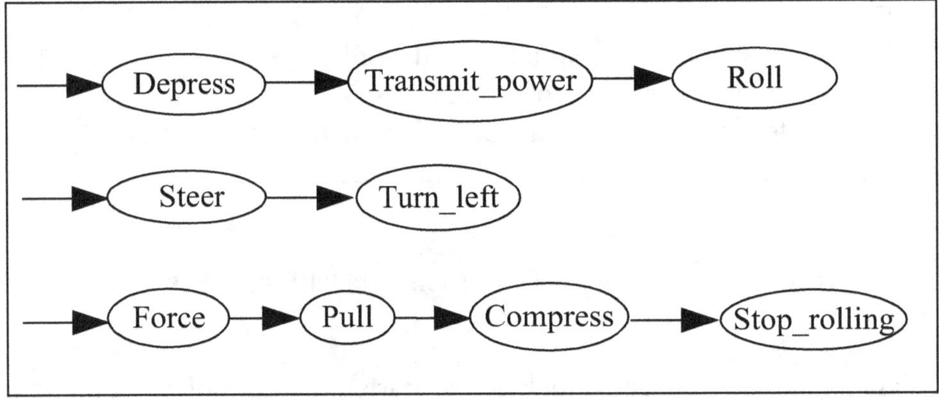

Figure 1-10 Functional Decomposition Method

1-4 Systems Definition 2.0

Systems structure and systems behavior are the two most significant views of a system. In order to achieve a truly integrated whole of a system, we first need to integrate the systems structure and systems behavior together. In other words, integration of systems structure and systems behavior results in the integration of a whole system. Since systems definition 1.0 does not define the integration of systems structure and systems behavior, very likely it only hopes and will never be able to really form an integrated whole of a system. In this situation, systems definition 1.0 is powerless in defining a system appropriately.

Structure-behavior coalescence (SBC) provides an elegant way to integrate the systems structure and systems behavior of a system. A system is therefore re-described, by systems definition 2.0, truly to be an integrated whole, through structure-behavior coalescence, embodied in its assembled components, their interactions with each other and the environment as shown in Figure 1-11.

A system,
through the SBC approach,
truly is an integrated whole,
embodied in its assembled components,
their interactions with each other and the environment.

Figure 1-11 Systems Definition 2.0 Defining a System

Systems definition or system definition 2.0 formally defines the essence of a system and its details at the same time. Since systems definition 2.0 demands the integration of systems structure and systems behavior, definitely it is able to form an integrated whole of a system. In this situation, systems definition 2.0 is fully capable of defining a system.

A system defined by systems definition 2.0 has the following characteristics: 1) it emphasizes the system's structure-behavior coalescence; 2) it is a truly integrated whole; 3) it is embodied in its assembled components; 4) components are interacting (or handshaking) with each other and the environment; and 5) it uses structural decomposition [Chao14a, Chao14b, Chao14c, Ghar11] rather than functional decomposition [Scho10].

Chapter 2: Introduction to Process Algebra

In this chapter, we first discuss what process algebra is. Then we show several examples of process algebras. After that we elaborate on the generalized SBC process algebra and six specialized SBC process algebras.

2-1 What is Process Algebra?

Process algebras are a diverse family of related approaches to the study of concurrent systems [Berg87, Chao15d, Hoar85, Miln89, Miln99]. Their tools are algebraic languages for the high-level description of interactions, communications and synchronizations between a collection of independent agents or processes.

Process algebras also provide algebraic laws that allow process descriptions to be manipulated and analyzed, and permit formal reasoning about equivalences and observation congruence among processes.

2-2 Examples of Process Algebras

There are several leading algebraic approaches to modeling concurrent systems.

Communicating Sequential Processes (CSP) [Hoar85] was first described in a 1978 paper by C. A. R. Hoare.

Arthur John Robin Gorell Milner introduced the Calculus of Communicating Systems (CCS) [Miln89, Miln99] around 1980.

Algebra of Communicating Processes (ACP) [Berg87] was initially developed by Jan Bergstra and Jan Willem Klop in 1982.

2-3 Generalized SBC Process Algebra

Generalized SBC process algebra (G-SBC-PA) evolved from CCS (Calculus of Communicating Systems) [Miln89, Miln99].

CCS is a general process algebra language for the study of communication and concurrency. Like CCS, generalized SBC process algebra is also a general process algebra language for the study of communication and concurrency.

2-4 Specialized SBC Process Algebras

Channel-based single-queue SBC process algebra (C-S-SBC-PA) [Chao15d, Chao15e, Chao15g], channel-based multi-queue SBC process algebra (C-M-SBC-PA) [Chao15d, Chao15f, Chao15h], channel-based infinite-queue SBC process algebra (C-I-SBC-PA) [Chao15b, Chao15c, Chao15d], operation-based single-queue SBC process algebra (O-S-SBC-PA), operation-based multi-queue SBC process algebra (O-M-SBC-PA) [Chao15d, Chao15f, Chao15h] and operation-based infinite-queue SBC process algebra (O-I-SBC-PA) [Chao15b, Chao15c, Chao15d] are the six specialized SBC process algebras.

Channel-based single-queue SBC process algebra, channel-based multi-queue SBC process algebra, channel-based infinite-queue SBC process algebra, operation-based single-queue SBC process algebra, operation-based multi-queue SBC process algebra and operation-based infinite-queue SBC process algebra all evolved from CCS (Calculus of Communicating Systems) [Miln89, Miln99].

CCS is a general process algebra language for the study of concurrent systems. Unlike CCS, six specialized SBC process algebras are only applicable to systems definition [Burd10, Maie09, Chao16b, Chao16c, Chao16d, Chao16e, Chao16f, Chao16g, Craw15, Dam06, O'Rou03, Putm00, Rayn09, Roza11, Toga08].

Chapter 3: Mathematics of SBC Process Algebra

To give the SBC process a mathematical definition, we need a means to form new processes from old ones. The basic operators, always present in some form or other, allow sequentialization of prefixes or summation of processes or parallel composition of processes or recursive definition of a process or replication of processes or conditional definition of a process or null process.

3-1 Sequentialization of Prefixes

Sometimes prefixes must be temporally ordered. For example, it might be desirable to specify algorithms such as: execute the "*t*" prefix (interaction with a condition) first and then execute the "*P*" process later. Sequentialization of prefixes can be used for such purposes.

Sequentialization of prefixes, usually written as the $t \bullet P$ process, indicates that it will perform the "*t*" prefix first and continue as the "*P*" process.

3-2 Summation of Processes

The binary operator "+", summation, combines two process expressions as alternatives.

For example, the process $P+Q$ can proceed non-deterministically either as the process P or the process Q; as soon as one performs its first interaction the other is discarded.

3-3 Parallel Composition of Processes

Parallel composition of two processes P and Q, usually written $P||Q$, is the key primitive distinguishing the process algebras from sequential models of process executions.

Parallel composition allows the executions in P and Q to proceed simultaneously and independently.

3-4 Recursive Definition of a Process

The operators presented so far describe only finite interaction and are consequently insufficient for full computability, which includes non-terminating behavior. Recursion is the operator that allows finite descriptions of infinite behavior.

For example, **fix**(X=E) can be understood as abbreviating the recursive definition of an infinite behavior denoted by the "X" process variable.

3-5 Replication of a Process

Replication is the other operator that allows finite descriptions of infinite behavior of a process.

For example, replication !P can be understood as abbreviating the parallel composition of a countably infinite number of P processes.

3-6 Conditional Definition of a Process

A process can be defined by a one-or-more-armed conditional expression. For example, the process (**if** $cond_1$ **then** a_1)+(**if** $cond_2$ **then** a_2)…+(**if** $cond_j$ **then** a_j) will proceed to perform the "a_1" interaction if the "$cond_1$" value is true, or proceed to perform the "a_2" interaction if the "$cond_2$" value is true,…, or proceed to perform the "a_j" interaction if the "$cond_j$" value is true.

3-7 Null Process

Process algebras generally also include a null process, denoted as *STOP*, which has no interaction points. It is utterly inactive and its sole purpose is to act as the inductive anchor on top of which more interesting processes can be generated.

The process "$STOP \bullet P_1$" (i.e. sequential composition of processes *STOP* and P_1) equals to the process "*STOP*", as shown in Figure 3-1.

$$STOP \bullet P_1 \quad = \quad STOP$$

Figure 3-1 Characteristics of Null Process (I)

The process "P_2+*STOP*" (i.e. summation of processes P_2 and *STOP*) equals to the process "*STOP*+P_2" (i.e. summation of processes *STOP* and P_2) which equals to the process "P_2", as shown in Figure 3-2.

$$P_2 + STOP \quad = \quad STOP + P_2 \quad = \quad P_2$$

Figure 3-2 Characteristics of Null Process (II)

The process "$P_3 || STOP$" (i.e. parallel composition of processes P_3 and *STOP*) equals to the process "$STOP || P_3$" (i.e. parallel composition of processes *STOP* and P_3) which equals to the process "P_3", as shown in Figure 3-3.

$$P_3 \,||\, STOP \quad = \quad STOP \,||\, P_3 \quad = \quad P_3$$

Figure 3-3 Characteristics of Null Process (III)

PART II: CHANNEL-BASED MULTI-QUEUE SBC PROCESS ALGEBRA

Chapter 4: Channel-Based Value-Passing Interactions

In this chapter, we first introduce channels and channel-based interactions. We then introduce the formal description of a channel-based communication port, a channel-based action and a channel-based interaction.

4-1 Channels

Channels are a model for agent communication. An agent may provide many channels, as shown in Figure 4-1.

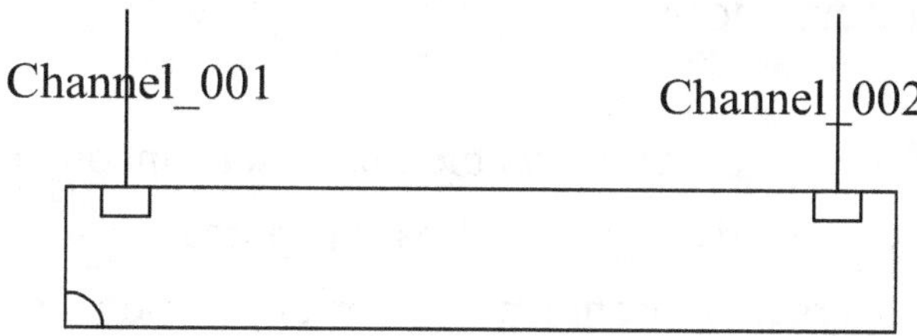

Figure 4-1. A An Agent May Provide Many Channels

A channel may contain several input parameters (e.g. i_1, i_2) and output parameters (e.g. o_1, o_2), as shown in Figure 4-2.

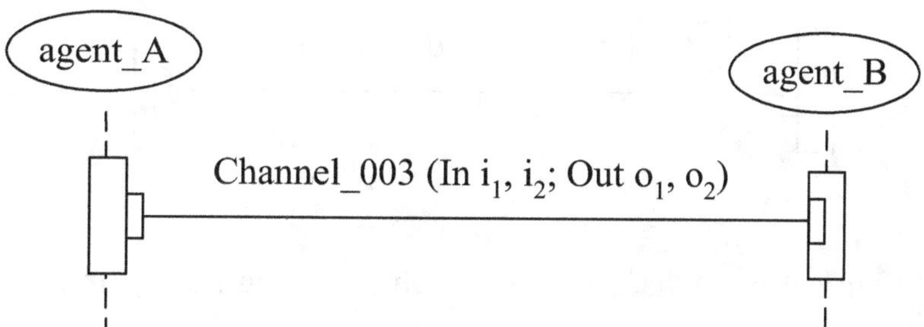

Figure 4-2. A Channel Contains Several Input/Output Parameters

A channel formula is used to completely describe a channel. A channel formula includes a) channel name, b) input parameters (e.g. $i_1, i_2, \ldots, i_m$) and c) output parameters (e.g. $o_1, o_2, \ldots, o_n$), as shown in Figure 4-3.

$$\text{Channel\_Name (In } i_1, i_2, \ldots, i_m \text{ ; Out } o_1, o_2, \ldots, o_n)$$

Figure 4-3 Channel Formula

4-2 Channel-Based Interactions

An interaction represents an indivisible and instantaneous communication or handshake between two agents. In the channel-based approach as shown in Figure 4-4, the caller agent (either external environment's actor or component) interacts with the callee agent (component)through the channel interaction.

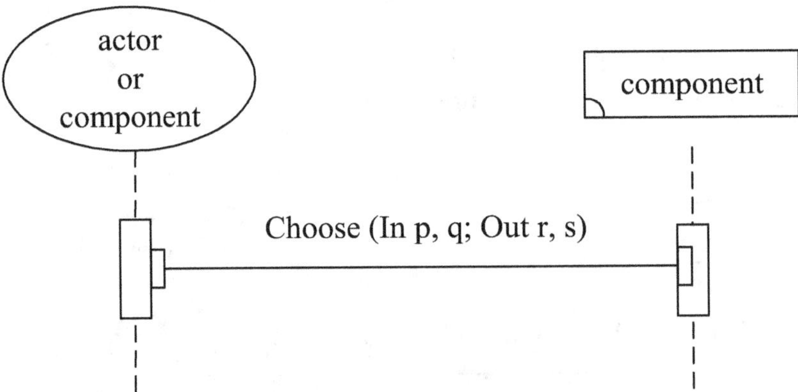

Figure 4-4 Channel-Based Value-Passing Interaction

The caller agent owns the "calling port" of the interaction. In this case, the calling port is " $\overline{\text{Choose (In p, q; Out r, s)}}$ " and its conduct is to assist the caller agent to output a value to each of the "p" and "q" variables (of the "Choose" channel), and input a value from each of the "r" and "s" variables (of the "Choose" channel), as shown in Figure 4-5.

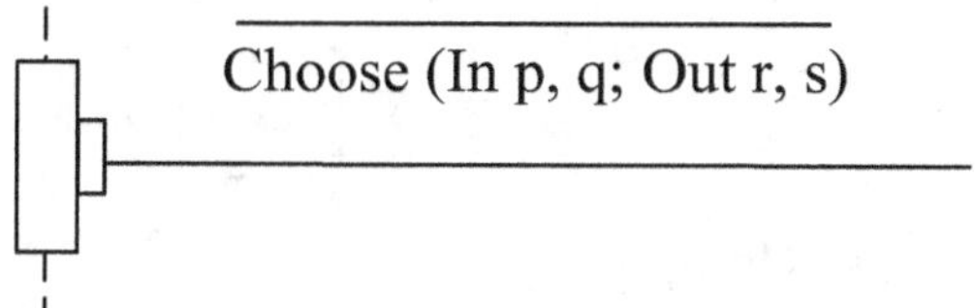

Figure 4-5 Calling Port

The caller agent together with the "calling port" is named the "calling action" as shown in Figure 4-6.

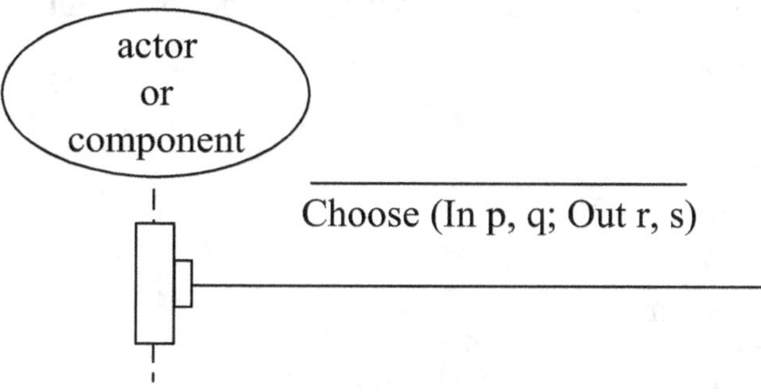

Figure 4-6 Calling Action

The callee agent owns the "called port" of the interaction. In this case, the called port is "Choose (In p, q; Out r, s)" and its conduct is to assist the callee agent to input a value from each of the "p" and "q" variables (of the "Choose" channel), and output a value to each of the "r" and "s" variables (of the "Choose" channel), as shown in Figure 4-7.

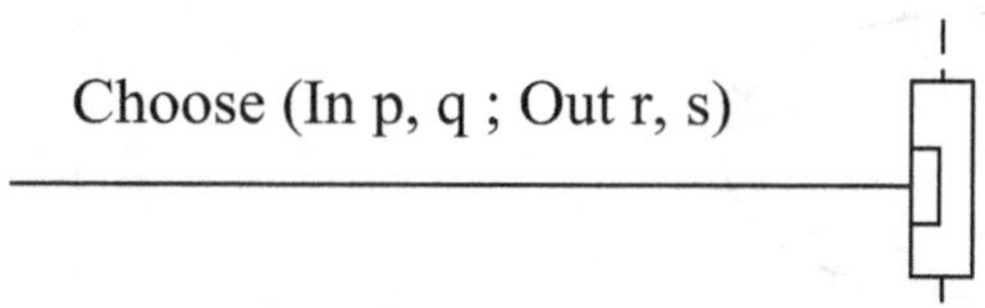

Figure 4-7 Called Port

The callee agent together with the "called port" is named the "called action" as shown in Figure 4-8.

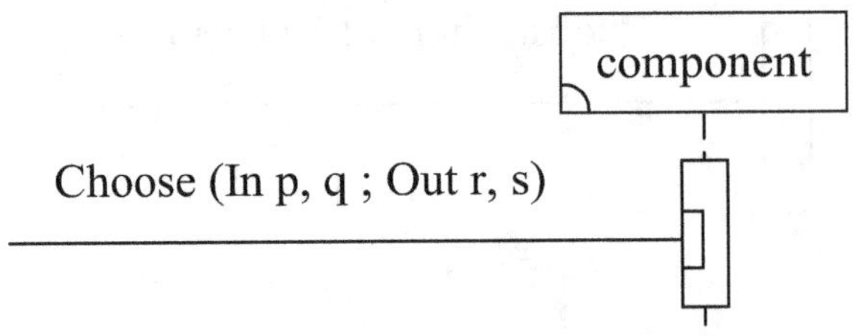

Figure 4-8 Called Action

In order to simplify the channel-based interaction diagram, we will redraw it as shown in Figure 4-9.

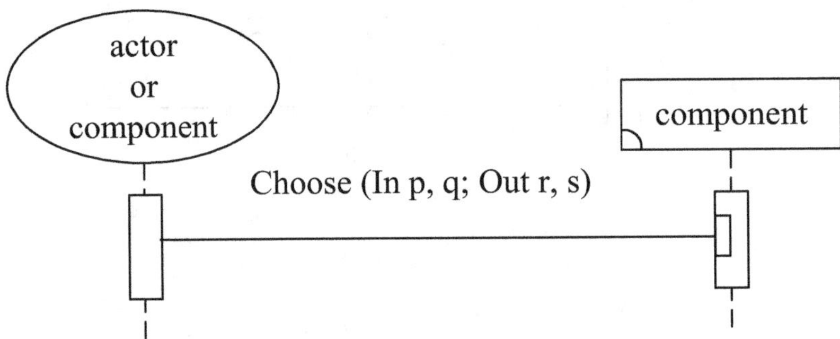

Figure 4-9 Channel-Based Interaction Diagram (I)

Or we can draw the channel-based interaction diagram as shown in Figure 4-10.

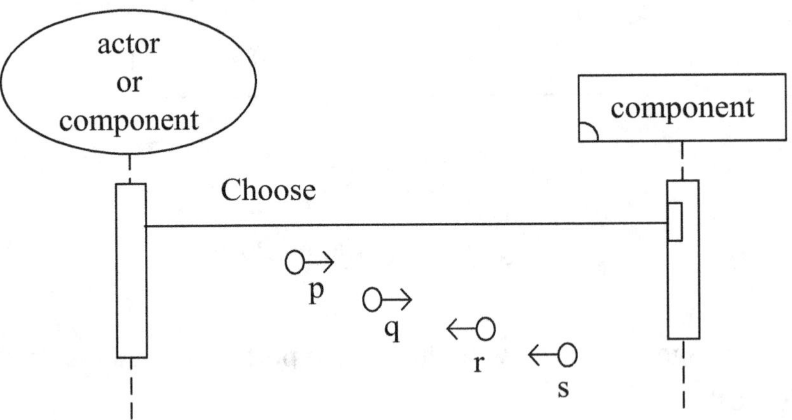

Figure 4-10 Channel-Based Interaction Diagram (II)

4-3 Formal Description of a Channel-Based Communication Port

We formally describe a channel-based communication port as a 2-tuple PORT = <calling_or_called, channel_formula>, where "calling_or_called" stands for a CALLING or CALLED port tag and "channel_formula" stands for a channel formula as shown in Figure 4-11.

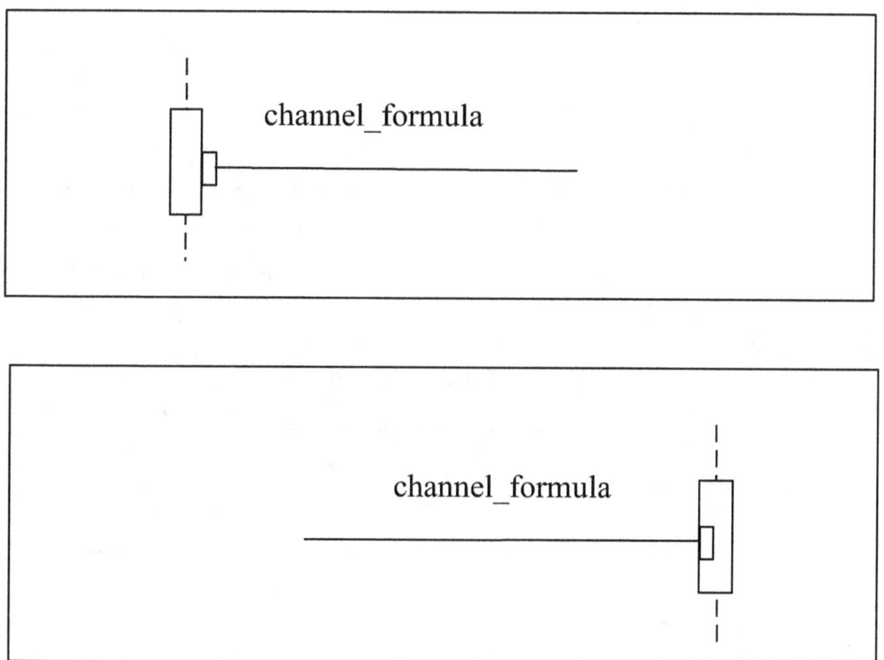

Figure 4-11 Formal Description of
a Channel-Based Communication Port

4-4 Formal Description of a Channel-Based Action

We formally describe a channel-based action as a 3-tuple ACTION = <agent, calling_or_called, channel_formula>, where "agent" stands for the name of a caller agent or callee component, "calling_or_called" stands for a CALLING or CALLED action tag, and "channel_formula" stands for a channel formula as shown in Figure 4-12.

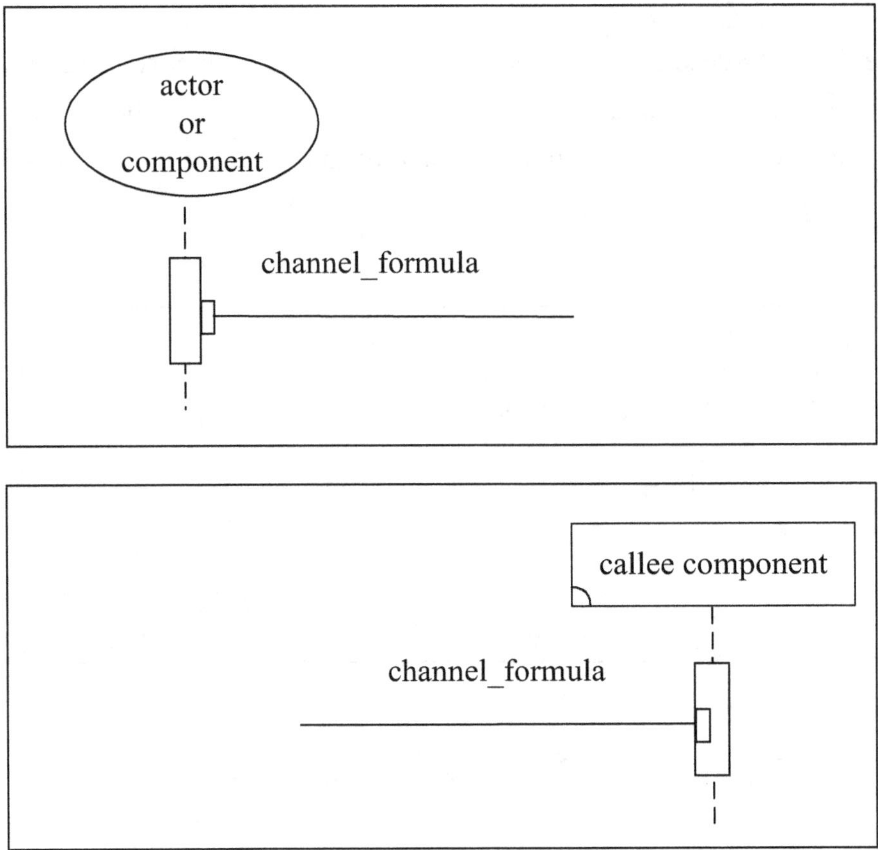

Figure 4-12 Formal Description of a Channel-Based Action

4-5 Formal Description of a Channel-Based Interaction

We formally describe a channel-based interaction as a 3-tuple INTERACTION = <caller_agent, channel_formula, callee_component>, where "caller_agent" stands for the name of a caller agent, "channel_formula" stands for a channel formula and "callee_component" stands for the name of a callee component as shown in Figure 4-13.

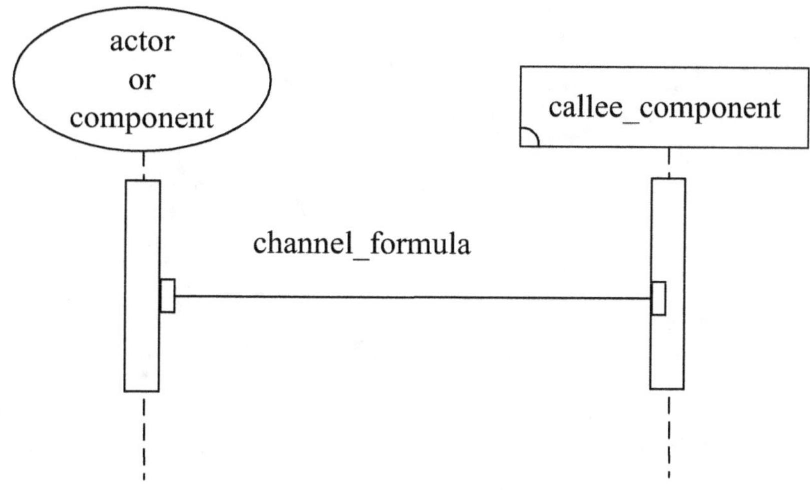

Figure 4-13 Formal Description of a Channel-Based Interaction

Chapter 5: The Structure-Behavior Coalescence Approach

Systems structure and systems behavior are the two most prominent views of a system, integrating the systems structure and systems behavior is apparently the best way to achieve an integrated whole of a system. If we are not able to integrate the systems structure and systems behavior, then there is no way that we are able to integrate the whole system. Structure-behavior coalescence (SBC) provides an elegant way to integrate the systems structure and systems behavior of a system. In other words, SBC facilitates an integrated whole of a system.

5-1 Structure-Behavior Coalescence Means to Integrate the Systems Structure and Systems Behavior

Systems structure, specified by components, their channels and their composition, refers to the type of connection between the components of a system as shown in Figure 5-1.

Figure 5-1 Systems Structure

Systems behavior, specified by the interactions between and among the components and environment, refers to the interconnectivities a system in conjunction with its environment as shown in Figure 5-2.

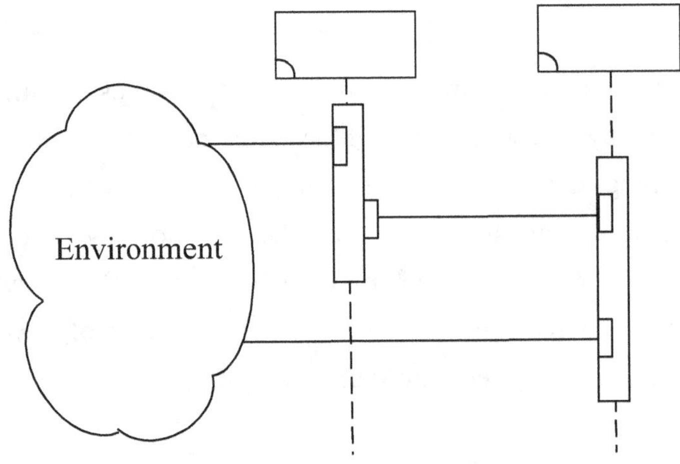

Figure 5-2 Systems Behavior

Systems structure and systems behavior are the two most prominent views of a system, integrating the systems structure and systems behavior is apparently the best way to achieve an integrated whole of a system.

If we are not able to integrate the systems structure and systems behavior, then there is no way that we are able to integrate the whole system.

Structure-behavior coalescence (SBC) [Chao15a] provides an elegant way to integrate the systems structure and systems behavior of a system. In other words, SBC facilitates an integrated whole of a system.

5-2 Interactions among Components and the External Environment to Draw Forth the Systems Behavior

All things that strike us as something independent are essentially parts of a system. We usually call the parts of a system its components. Components are sometimes labeled as parts, entities, objects, building blocks and non-aggregated systems [Chao14a, Chao14b, Chao14c, Chao16a].

In a system, if the components, and among them and the external environment to interact (or handshake), such interaction will draw forth the systems behavior.

The external environment uses a “type-1 interaction” to interact with a component as shown in Figure 5-3.

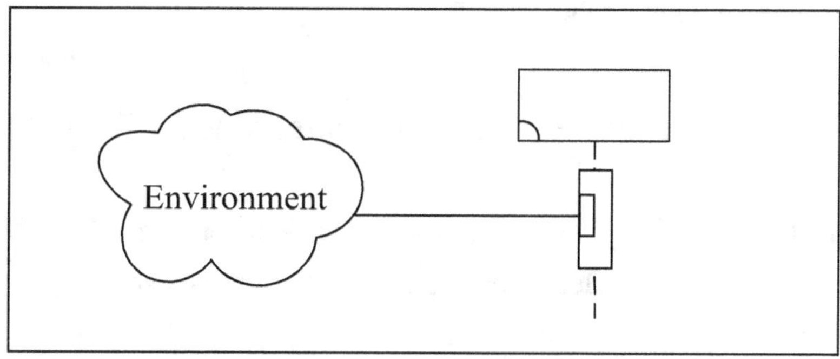

Figure 5-3 The External Environment Uses a "Type-1 Interction" to Interact with a Component

We formally describe a channel-based "type-1 interaction" as a 3-tuple TYPE_1_INTERACTION = <actor, channel_formula, callee_component>, where "actor" stands for the name of an external environment's actor, "channel_formula" stands for a channel formula and "callee_component" stands for the name of a callee component.

Two components use a "type-2 interaction" to interact with each other as shown in Figure 5-4.

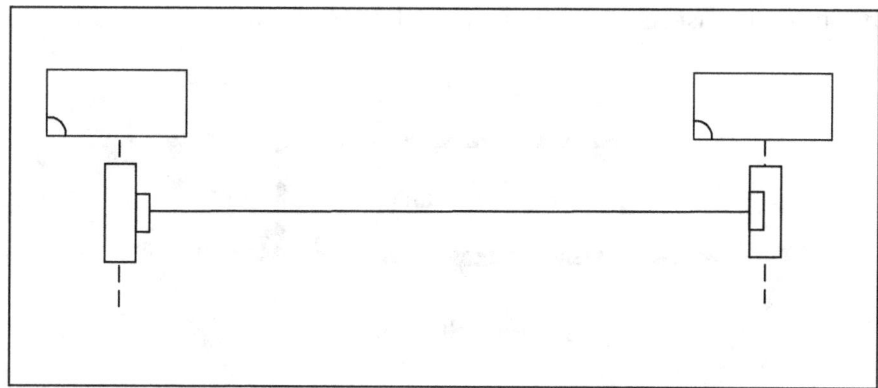

Figure 5-4 Two Components Use a "Type_2 Interaction" to Interact with Each Other

We formally describe a channel-based "type-2 interaction" as a 3-tuple TYPE_2_INTERACTION = <caller_component, channel_formula, callee_component>, where "caller_component" stands for the name of a caller component, "channel_formula" stands for a channel formula and "callee_component" stands for the name of a callee component.

5-3 Core Theme of Structure-Behavior Coalescence

The core theme of structure-behavior coalescence is: "Systems Architecture = Systems Structure + Systems Behavior." That is, the systems structure will lead to the systems behavior as shown in Figure 5-5.

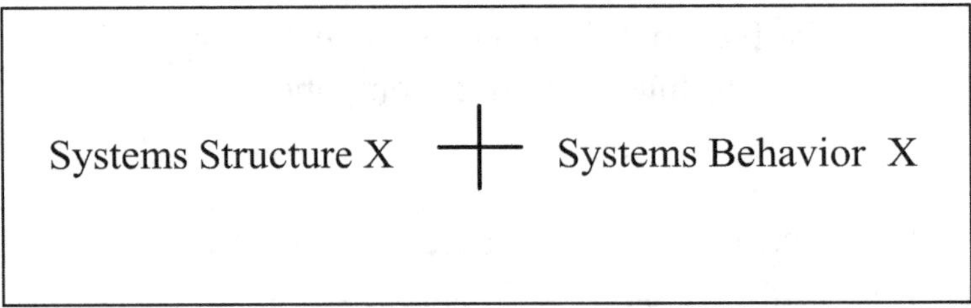

Figure 5-5 Core Theme of Structure-Behavior Coalescence

One systems structure will draw forth one systems behavior. That is, the systems behavior is attached to or built on the systems structure in the SBC approach.

In other words, the systems behavior can not exist alone; it must be loaded on the systems structure just like a cargo is loaded on a ship as shown in Figure 5-6.

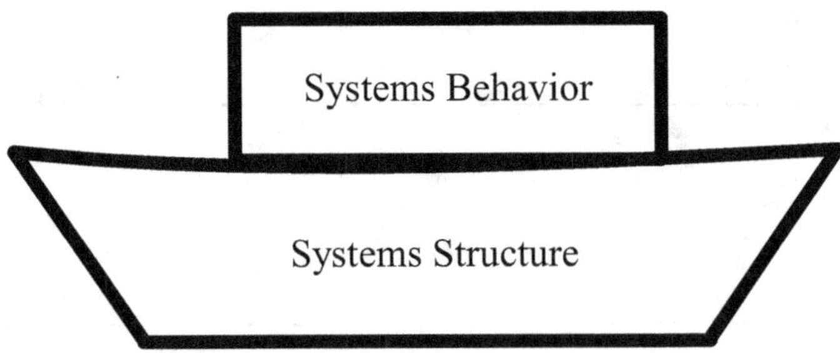

Figure 5-6 Systems Behavior Must be Loaded on the Systems Structure

Chapter 6: Language Constructs of Channel-Based Multi-Queue SBC Process Algebra

In the chapter, we illustrate in detail those channel-based multi-queue SBC process algebra language constructs which make up the system definition of a system.

6-1 Entity Set and Entity Name

As shown in Figure 6-1, we assume an infinite set K of channel formulas, and use k_1, k_2...to range over K. Further, we let N be the set of condition type_1 interactions, and use n_1, n_2...to range over N. We let Z be the set of condition type_2 interactions, and use z_1, z_2...to range over Z. We let T be the set of condition type_1_or_2 interactions, and use t_1, t_2...to range over T. Further, we let X be the set of process variables, and use X_1, X_2...to range over X. We let Φ be the set of process Constants, and use A_1, A_2...to range over Φ. We let Π be the set of processes, and use P_1, Q_1...to range over Π. We let Ψ be the set of process expressions, and use E_1, E_2...to range over Ψ. Finally, we let Γ be the set of components, and use C_1, C_2...to range over Γ.

Entity set	Entity name	Type of entity
K	k_1, k_2...	channel formulas
N	n_1, n_2...	condition type_1 interactions
Z	z_1, z_2...	condition type_2 interactions
T	t_1, t_2...	condition type_1_or_2 interactions
Δ	a_1, a_2...	type 1 or 2 interactions
X	X_1, X_2...	process variables
Φ	A_1, A_2...	process Constants
	I, J,...	indexing sets
Π	P_1, Q_1...	processes
Ψ	E_1, E_2...	process expressions
Γ	C_1, C_2...	components

Figure 6-1 Entities

6-2 Examples of Entities

As a first example, the n_{001} prefix defined as "<$Actor_{001}$, k_{001}, C_{001}>" or "**if** *true* **then** <$Actor_{001}$, k_{001}, C_{001}>" is a channel-based condition type_1 interaction under a tautology condition in which "$Actor_{001}$" stands for the external environment, "k_{001}" stands for a channel formula, and "C_{001}" stands for a callee component as shown in Figure 6-2.

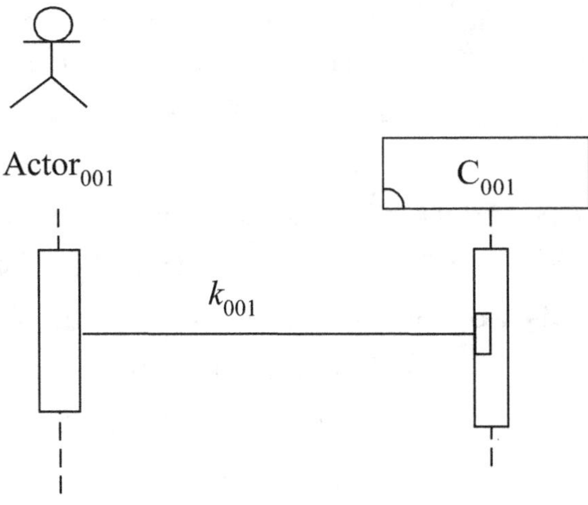

Figure 6-2 Prefix of n_{001}

As a second example, the n_{002} prefix defined as "**if** $cond_{002}$ **then** <$Actor_{002}$, k_{002}, C_{002}>" is a channel-based condition type_1 interaction under a certain (tautology excluded) condition in which "$cond_{002}$" stands for a (tautology excluded) condition in which "$Actor_{002}$" stands for the external environment, "k_{002}" stands for a channel formula, and "C_{002}" stands for a callee component as shown in Figure 6-3.

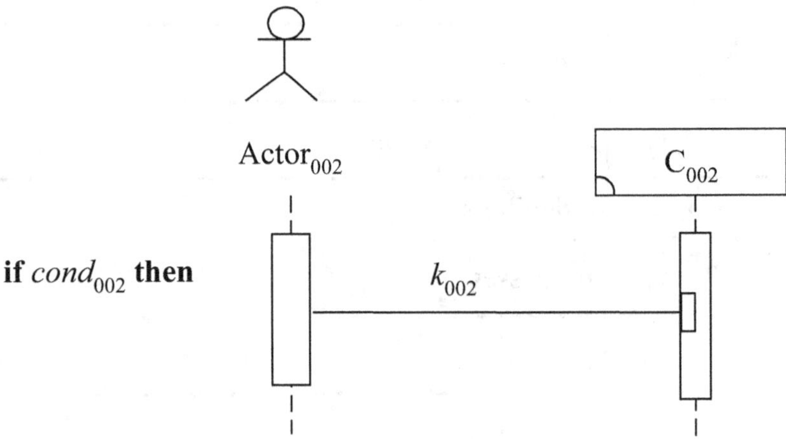

Figure 6-3 Prefix of n_{002}

As a third example, the z_{003} prefix defined as "<C_{003a}, k_{003}, C_{003b}>" or "**if** *true* **then** <C_{003a}, k_{003}, C_{003b}>" is a channel-based condition type_2 interaction under a tautology condition in which "C_{003a}" stands for a caller component, "k_{003}" stands for a channel formula, and "C_{003b}" stands for a callee component as shown in Figure 6-4.

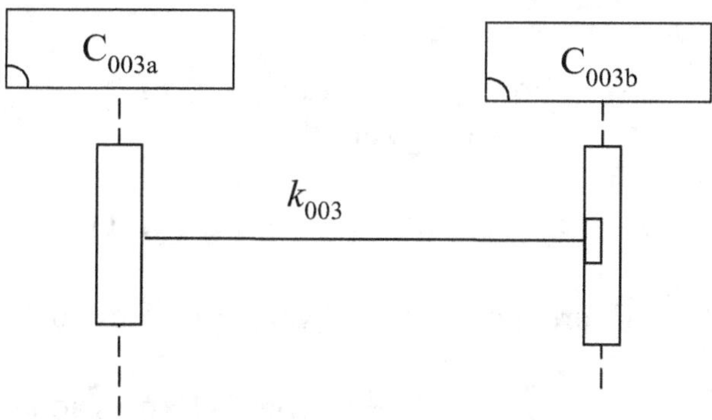

Figure 6-4 Prefix of z_{003}

As a fourth example, the z_{004} prefix defined as "**if** $cond_{004}$ **then** <C_{004a}, k_{004}, C_{004b}>" is a channel-based condition type_2 interaction under a certain (tautology excluded) condition in which "$cond_{004}$" stands for a (tautology excluded) condition in which "C_{004a}" stands for a caller component, "k_{004}" stands for a channel formula, and "C_{004b}" stands for a callee component as shown in Figure 6-5.

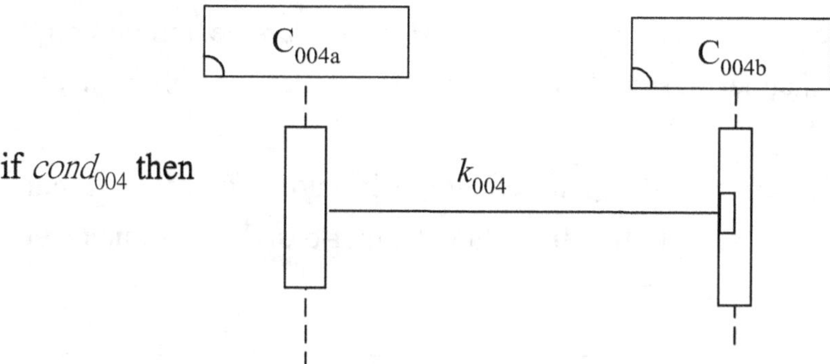

Figure 6-5 Prefix of z_{004}

6-3 Backus-Naur Form of Channel-Based Multi-Queue SBC Processes

The set of channel-based multi-queue SBC (i.e. Structure-Behavior Coalescence) processes is defined by the following BNF grammar, as shown in Figure 6-6.

(1) <System> ::= <FixIFD> {"|| " <FixIFD>}

(2) <FixIFD> ::= "fix(" <Process_Variable>"="<IFD>
"● " <Process_Variable> ")"

(3) <IFD> ::= <Type_1_Expression> {"● " <Type_1_Or_2_Expression>}

(4) <Type_1_Or_2_Expression> ::= <Type_1_Expression>

| <Type_2_Expression>

(5) <Type_1_Expression> ::= <Condition_Type_1_Interaction>
{"+" <Condition_Type_1_Interaction>}

(6) <Type_2_Expression> ::= <Condition_Type_2_Interaction>
{"+" <Condition_Type_2_Interaction>}

Figure 6-6 Backus-Naur Form of
Channel-Based Multi-Queue SBC Processes

6-3-1 Parallel Composition of One or More Recursive Interaction Flow Diagrams Defines the Channel-Based Multi-Queue SBC Process of a System

Rule 1 describes that parallel composition of one or more recursive interaction flow diagrams (i.e. FixIFD) defines the channel-based multi-queue SBC process of a system, as shown in Figure 6-7.

Rule 1
<System> ::= <FixIFD> {"|| " <FixIFD>}

Figure 6-7 Rule 1

6-3-2 A Recursive Interaction Flow Diagram is the Recursion of an Interaction Flow Diagram

Rule 2 describes that a recursive interaction flow diagram (i.e. FixIFD) is defined by the recursion of an interaction flow diagram (i.e. IFD), as shown in Figure 6-8.

Rule 2
<FixIFD> ::= "**fix**(" <Process_Variable> "=" <IFD> " ● " <Process_Variable> ")"

Figure 6-8 Rule 2

6-3-3 An Interaction Flow Diagram is a Type_1 Expression Followed by Zero or More Type_1_Or_Type_2 Expressions

Rule 3 describes that an interaction flow diagram (i.e. IFD) is defined by a type_1 expression (i.e. Type_1_Expression) followed by zero or more type_1_or_type_2 expressions (i.e. Type_1_Or_2_Expression), as shown in Figure 6-9.

Rule 3
<IFD> ::= <Type_1_Expression> {"●" <Type_1_Or_2_Expression>}

Figure 6-9 Rule 3

6-3-4 Type_1_Or_2 Expression is either Type_1 or Type_2

Rule 4 describes that the type_1_or_2 expression (i.e. Type_1_Or_2_Expression) is either a type_1 expression (i.e. Type_1_Expression) or a type_2 expression (i.e. Type_2_Expression), as shown in Figure 6-10.

Rule 4
<Type_1_Or_2_Expression> ::= <Type_1_Expression> | <Type_2_Expression>

Figure 6-10 Rule 4

6-3-5 Summation of One or More Condition Type_1 Interactions Defines the Type_1 Expression

Rule 5 describes that the summation of one or more condition type_1 interactions (i.e. Condition_Type_1_Interaction) defines the type_1 expression (i.e. Type_1_Expression), as shown in Figure 6-11.

Rule 5
<Type_1_Expression> ::= <Condition_Type_1_Interaction> {"+" <Condition_Type_1_Interaction>}

Figure 6-11 Rule 5

6-3-6 Summation of One or More Condition Type_2 Interactions Defines the Type_2 Expression

Rule 6 describes that the summation of one or more condition type_2 interactions (i.e. Condition_Type_2_Interaction) defines the type_2 expression (i.e. Type_1_Expression), as shown in Figure 6-12.

Rule 6
<Type_2_Expression> ::= <Condition_Type_2_Interaction> {"+" <Condition_Type_2_Interaction>}

Figure 6-12 Rule 6

6-4 Examples of Type_1 Expression

As a first example, consider the type-1 expression E_{011} is defined as "n_{011}" and the n_{011} prefix is defined as "<$Actor_{011}$, k_{011}, C_{011}>". The Backus-Naur Form tree of E_{011} is shown in Figure 6-13.

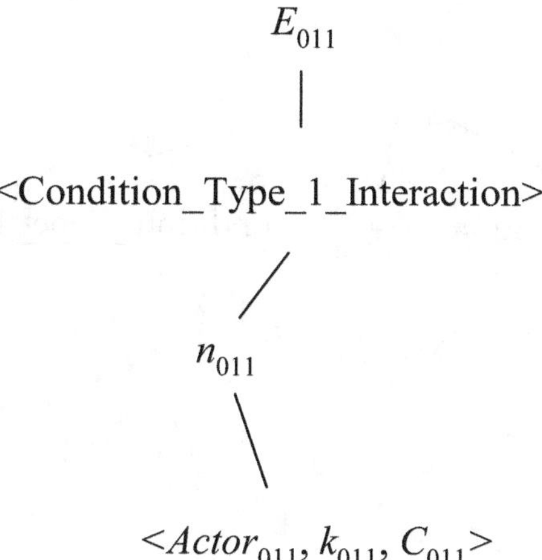

Figure 6-13 Backus-Naur Form Tree of the E_{011} Type-1 Expression

As a second example, consider the type-1 expression E_{012} is defined as "n_{012}" and the n_{012} prefix is defined as "**if** $cond_{012}$ **then** <$Actor_{012}$, k_{012}, C_{012}>". The Backus-Naur Form tree of E_{012} is shown in Figure 6-14.

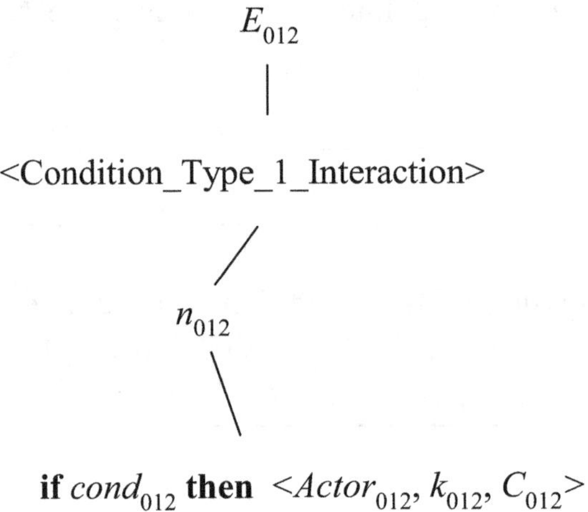

Figure 6-14 Backus-Naur Form Tree of the E_{012} Type-1 Expression

As a third example, consider the type-1 expression E_{013} is defined as "$n_{013a}+n_{013b}$" and the n_{013a}, n_{013b} prefixes are defined respectively as "<$Actor_{013a}$, k_{013a}, C_{013a}>", "<$Actor_{013b}$, k_{013b}, C_{013b}>". The Backus-Naur Form tree of E_{013} is shown in Figure 6-15.

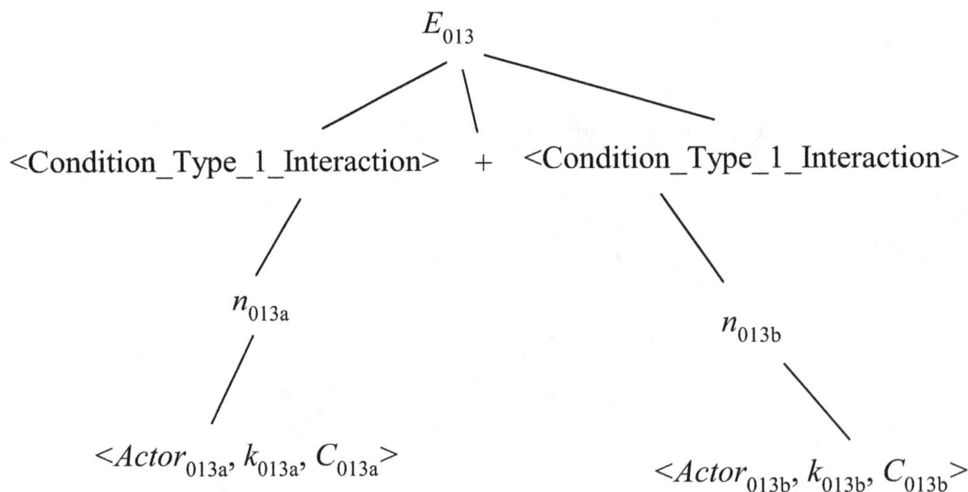

Figure 6-15 Backus-Naur Form Tree of the E_{013} Type-1 Expression

As a fourth example, consider the type-1 expression E_{014} is defined as "$n_{014a}+n_{014b}$" and the n_{014a}, n_{014b} prefixes are defined respectively as "<$Actor_{014a}$, k_{014a}, C_{014a}>", "**if** $cond_{014b}$ **then** <$Actor_{014b}$, k_{014b}, C_{014b}>". The Backus-Naur Form tree of E_{014} is shown in Figure 6-16.

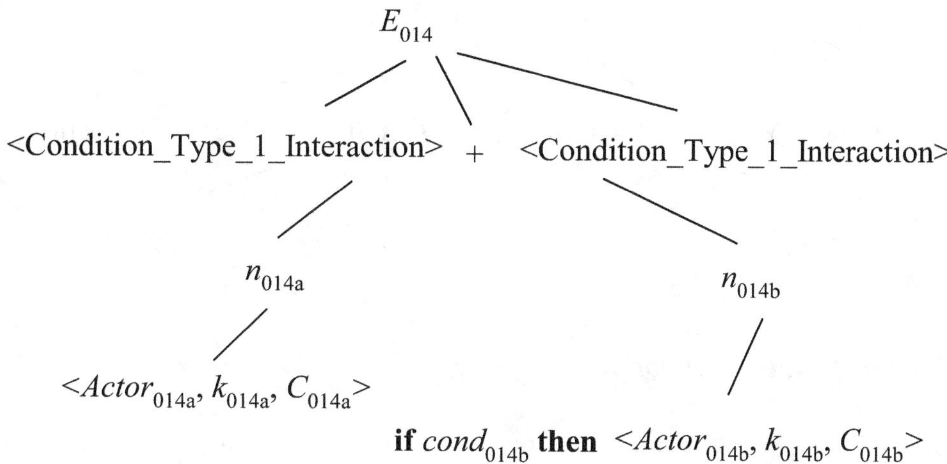

Figure 6-16 Backus-Naur Form Tree of the E_{014} Type-1 Expression

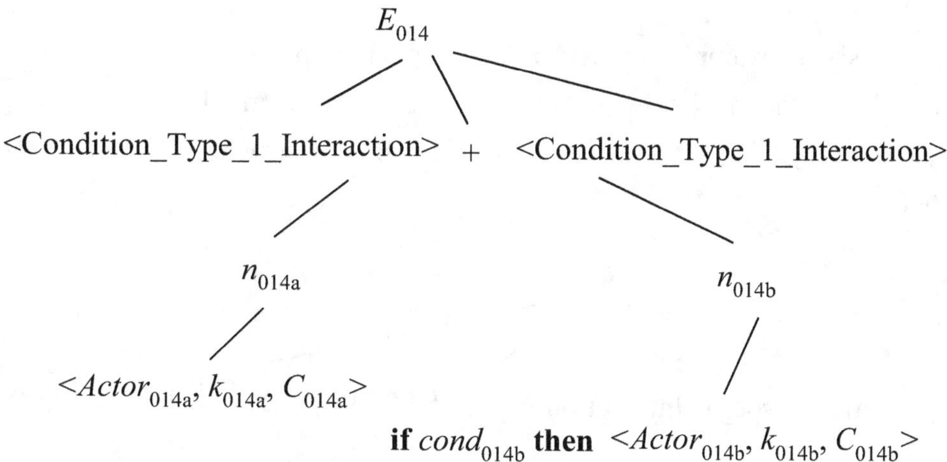

Figure 6-16 Backus-Naur Form Tree of the E_{014} Type-1 Expression

As a fifth example, consider the type-1 expression E_{015} is defined as "n_{015a}+n_{015b}" and the n_{015a}, n_{015b} prefixes are defined respectively as "**if** $cond_{015a}$ **then** <$Actor_{015a}$, k_{015a}, C_{015a}>", "**if** $cond_{015b}$ **then** <$Actor_{015b}$, k_{015b}, C_{015b}>". The Backus-Naur Form tree of E_{015} is shown in Figure 6-17.

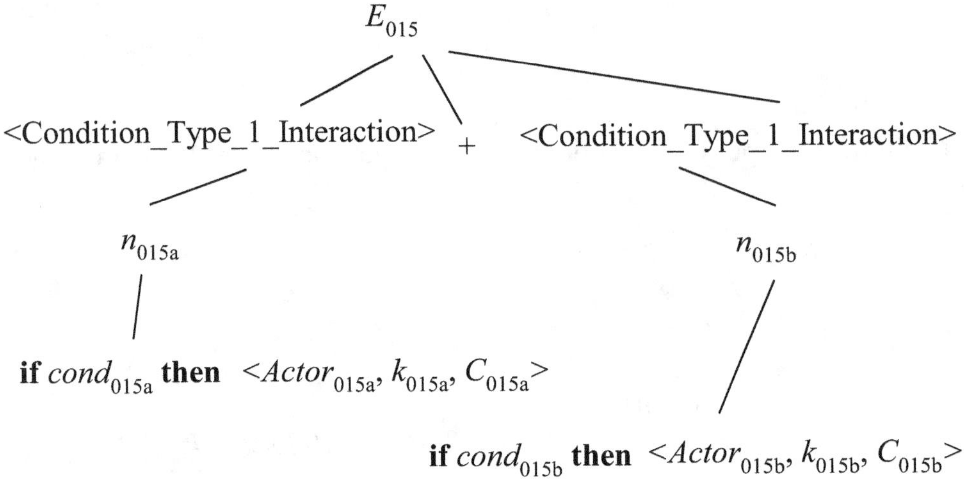

Figure 6-17 Backus-Naur Form Tree of the E_{015} Type-1 Expression

As a sixth example, consider the type-1 expression E_{016} is defined as "n_{016a}+n_{016b}" and the n_{013a}, n_{013b} prefixes are defined respectively as "<$Actor_{016}$, k_{016a}, C_{016a}>", "<$Actor_{016}$, k_{016b}, C_{016b}>". The Backus-Naur Form tree of E_{016} is shown in Figure 6-18.

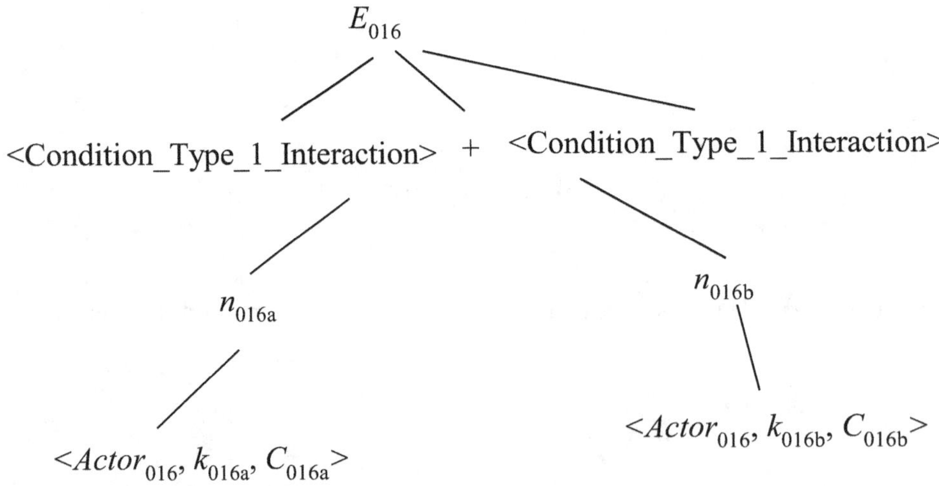

Figure 6-18 Backus-Naur Form Tree of the E_{016} Type-1 Expression

As a seventh example, consider the type-1 expression E_{017} is defined as "$n_{017a}+n_{017b}$" and the n_{017a}, n_{017b} prefixes are defined respectively as "<$Actor_{017}$, k_{017a}, C_{017a}>", "**if** $cond_{017b}$ **then** <$Actor_{017}$, k_{017b}, C_{017b}>". The Backus-Naur Form tree of E_{017} is shown in Figure 6-19.

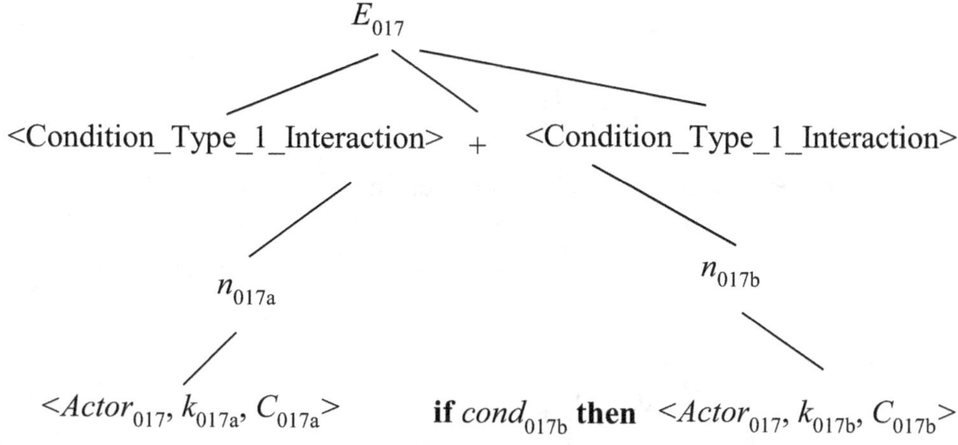

Figure 6-19 Backus-Naur Form Tree of the E_{017} Type-1 Expression

As an eighth example, consider the type-1 expression E_{018} is defined as "$n_{018a}+n_{018b}$" and the n_{018a}, n_{018b} prefixes are defined respectively as "**if** $cond_{018a}$ **then** <$Actor_{018a}$, k_{018a}, C_{018a}>", "**if** $cond_{018b}$ **then** <$Actor_{018b}$, k_{018b}, C_{018b}>". The Backus-Naur Form tree of E_{018} is shown in Figure 6-20.

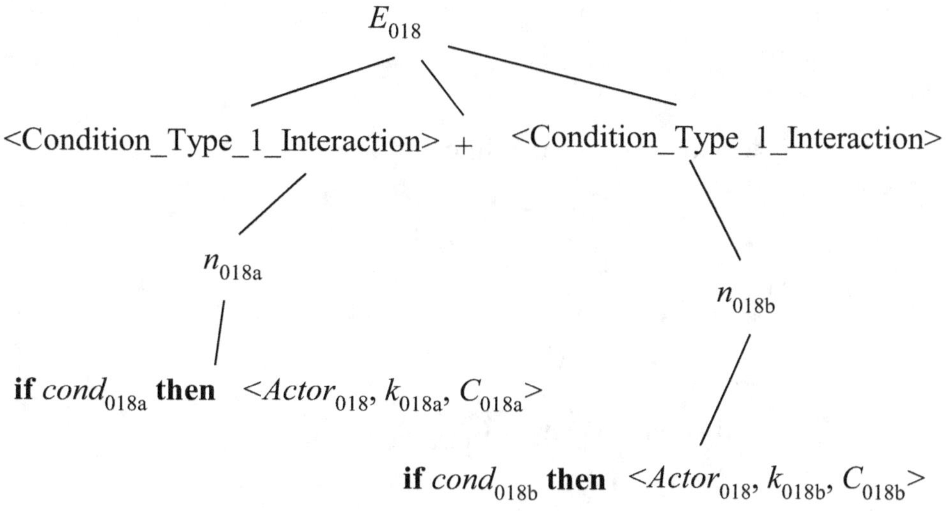

Figure 6-20 Backus-Naur Form Tree of the E_{018} Type-1 Expression

6-5 Examples of Type_2 Expression

As a first example, consider the type-2 expression E_{031} is defined as “z_{031}” and the z_{031} prefix is defined as “$<C_{031a}, k_{031}, C_{031b}>$”. The Backus-Naur Form tree of E_{031} is shown in Figure 6-21.

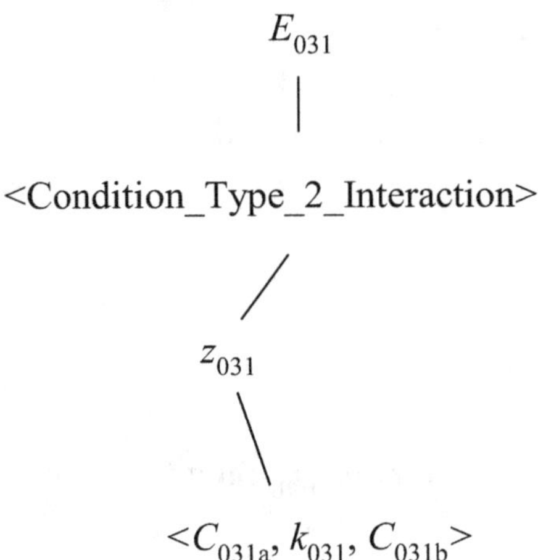

Figure 6-21 Backus-Naur Form Tree of the E_{031} Type-2 Expression

As a second example, consider the type-2 expression E_{032} is defined as “z_{032}” and the z_{032} prefix is defined as “**if** $cond_{032}$ **then** $<C_{032a}, k_{032}, C_{032b}>$”. The Backus-Naur Form tree of E_{032} is shown in Figure 6-22.

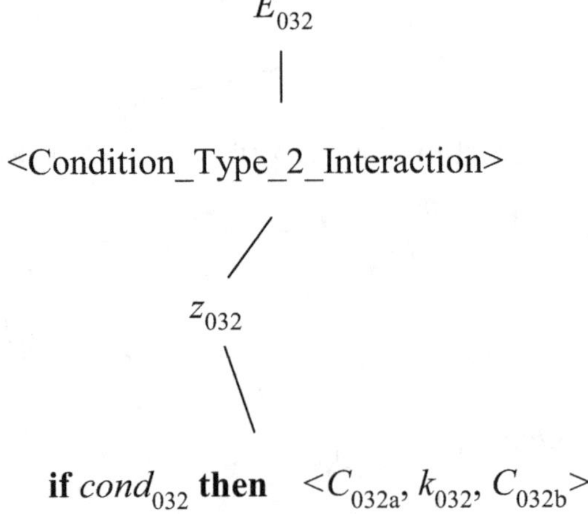

Figure 6-22 Backus-Naur Form Tree of the E_{032} Type-2 Expression

As a third example, consider the type-2 expression E_{033} is defined as "z_{033a}+z_{033b}" and the z_{033a}, z_{033b} prefixes are defined respectively as "$<C_{033a}, k_{033a}, C_{033c}>$", "$<C_{033b}, k_{033b}, C_{033d}>$". The Backus-Naur Form tree of E_{033} is shown in Figure 6-23.

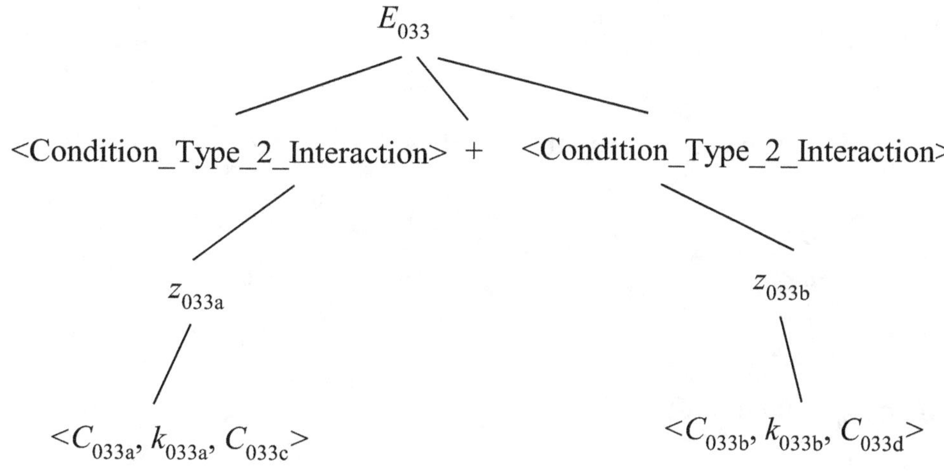

Figure 6-23 Backus-Naur Form Tree of the E_{033} Type-2 Expression

As a fourth example, consider the type-2 expression E_{034} is defined as "z_{034a}+z_{034b}" and the z_{034a}, z_{034b} prefixes are defined respectively as "$<C_{034a}, k_{034a}, C_{034c}>$", "**if** *cond*$_{034b}$ **then** $<C_{034b}, k_{034b}, C_{034d}>$". The Backus-Naur Form tree of E_{034} is shown in Figure 6-24.

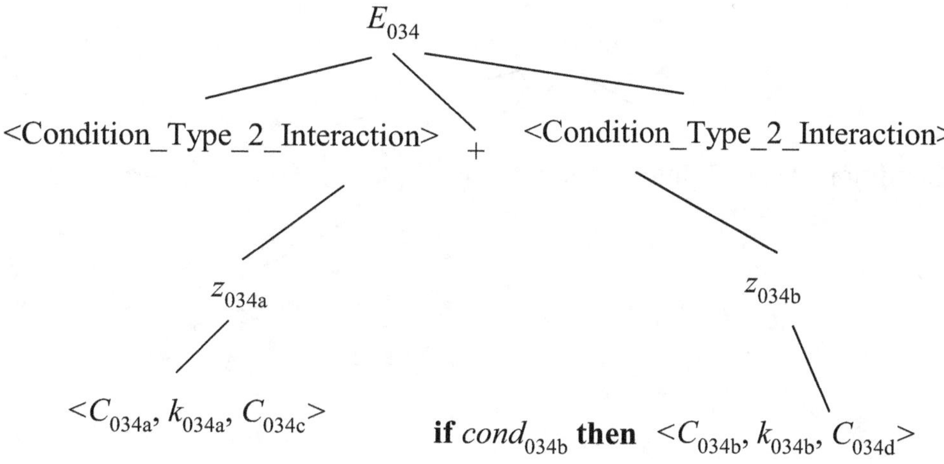

Figure 6-24 Backus-Naur Form Tree of the E_{034} Type-2 Expression

As a fifth example, consider the type-2 expression E_{035} is defined as "z_{035a}+z_{035b}" and the z_{035a}, z_{035b} prefixes are defined respectively as "**if** *cond*$_{035a}$ **then** <C_{035a}, k_{035a}, C_{035c}>", "**if** *cond*$_{035b}$ **then** <C_{035b}, k_{035b}, C_{035d}>". The Backus-Naur Form tree of E_{035} is shown in Figure 6-25.

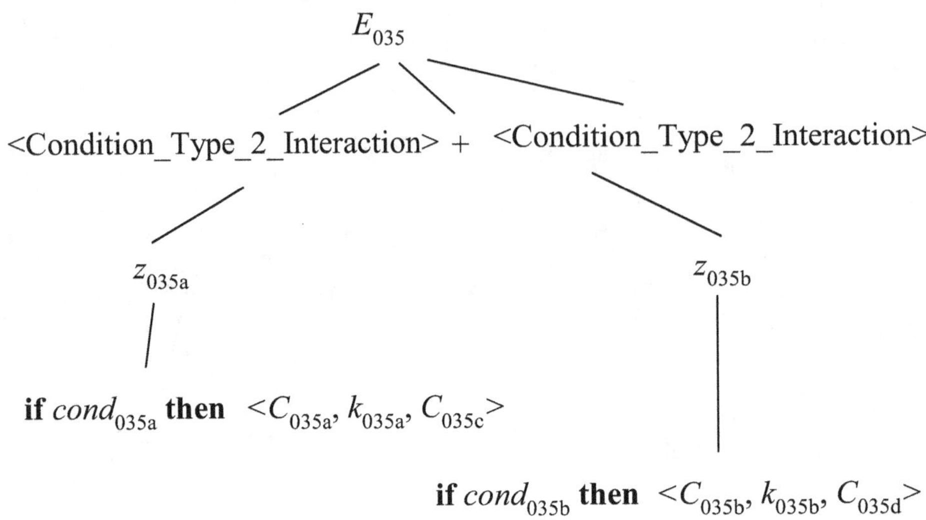

Figure 6-25 Backus-Naur Form Tree of the E_{035} Type-2 Expression

As a sixth example, consider the type-2 expression E_{036} is defined as "z_{036a}+z_{036b}" and the z_{036a}, z_{036b} prefixes are defined respectively as "<C_{036}, k_{036a}, C_{036a}>", "<C_{036}, k_{036b}, C_{036b}>". The Backus-Naur Form tree of E_{036} is shown in Figure 6-26.

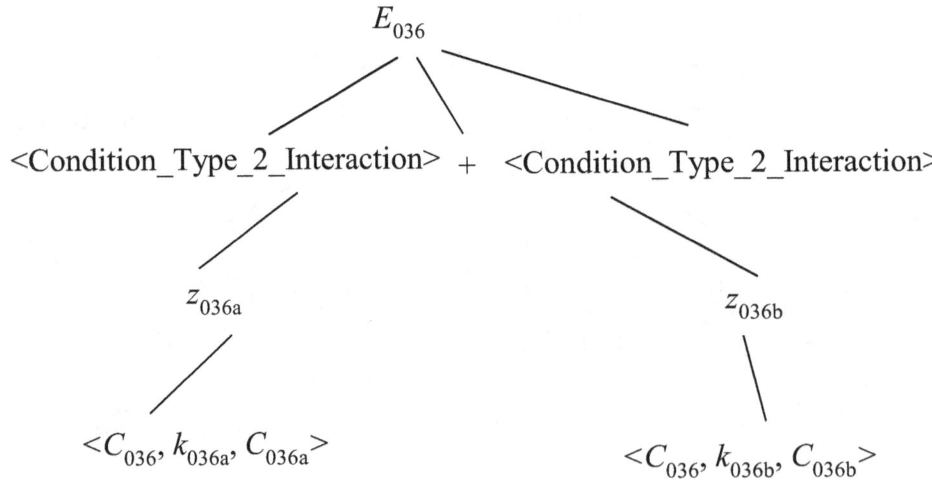

Figure 6-26 Backus-Naur Form Tree of the E_{036} Type-2 Expression

As a seventh example, consider the type-2 expression E_{037} is defined as "$z_{037a}+z_{037b}$" and the z_{037a}, z_{037b} prefixes are defined respectively as "$<C_{037}, k_{037a}, C_{037a}>$", "**if** $cond_{037b}$ **then** $<C_{037}, k_{037b}, C_{037b}>$". The Backus-Naur Form tree of E_{037} is shown in Figure 6-27.

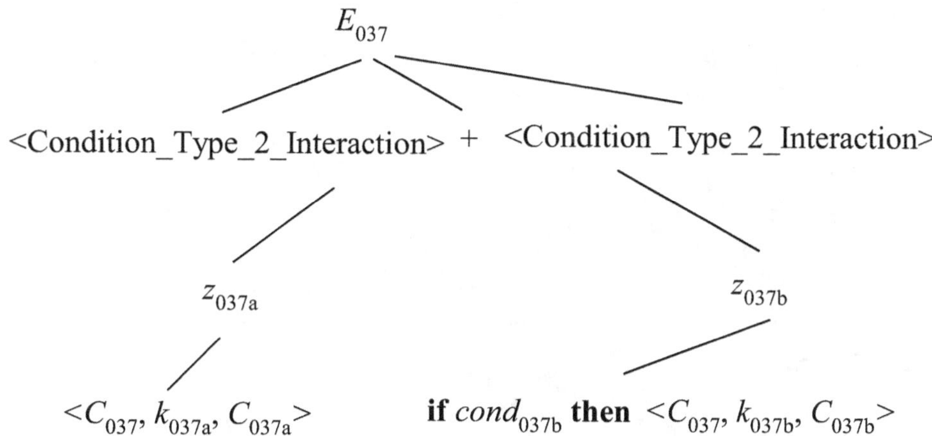

Figure 6-27 Backus-Naur Form Tree of the E_{037} Type-2 Expression

As an eighth example, consider the type-2 expression E_{038} is defined as "$z_{038a}+z_{038b}$" and the z_{038a}, z_{038b} prefixes are defined respectively as "**if** $cond_{038a}$ **then** $<C_{038}, k_{038a}, C_{038a}>$", "**if** $cond_{038b}$ **then** $<C_{038}, k_{038b}, C_{038b}>$". The Backus-Naur Form tree of E_{038} is shown in Figure 6-28.

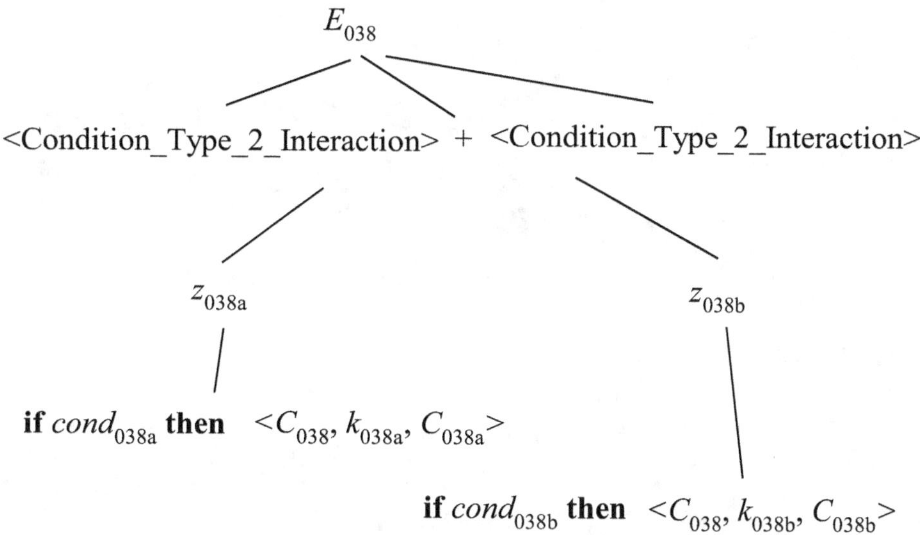

Figure 6-28 Backus-Naur Form Tree of the E_{038} Type-2 Expression

Chapter 7: Transitional Semantics of Channel-Based Multi-Queue SBC Process Algebra

In the chapter, we illustrate in detail those channel-based multi-queue SBC process algebra transitional semantics which regards the system definition of a system.

7-1 Transitional Semantics

In giving meaning to the channel-based multi-queue SBC process algebra, we shall use the following labelled transition system (LTS)

$$(\Psi, T, \rightarrow)$$

which consists of a set Ψ of process expressions, a set T of "condition type_1 or type_2 interactions", and a transition relation $\rightarrow \subseteq \Psi X T X \Psi$ where $(E_i, t, E_j) \in \rightarrow$ is denoted by $E_i \xrightarrow{t} E_j$.

The semantics for Ψ consists in the transition rules of each transition relation $\rightarrow$ over $\Psi X T X \Psi$. These transition rules will follow the structure of expressions.

As shown in Figure 7-1, we give the complete set of transition rules; the names Prefix, Sum, Parallel, Recursion, and Constant indicate that the rules are associated respectively with Prefix, Summation, Parallel Composition, and Recursion and with Constants.

Prefix	$$\frac{}{t \bullet E \xrightarrow{t} E}$$
Sum_j	$$\frac{E_j \xrightarrow{t} E'_j}{\sum_{i \in I} E_i \xrightarrow{t} E'_j} (j \in I)$$
Parallel_1	$$\frac{E \xrightarrow{t} E'}{E \parallel F \xrightarrow{t} E' \parallel F}$$
Parallel_2	$$\frac{F \xrightarrow{t} F'}{E \parallel F \xrightarrow{t} E \parallel F'}$$
Recursion	$$\frac{\mathbf{fix}(X=z\{\mathbf{fix}(X=z)/X\}) \xrightarrow{t} E'}{\mathbf{fix}(X=z) \xrightarrow{t} E'}$$
Constant	$$\frac{P \xrightarrow{t} P'}{A \xrightarrow{t} P'} \quad (A \overset{\text{def}}{=} P)$$

Figure 7-1 Transition Rules for the
Channel-Based Multi-Queue SBC Process Algebra

We said that the set of rules shown in Figure 7-1 is complete; by this we mean that there are no transitions except those which can be inferred by the rules.

7-2 Rule of Prefix

The rule for Prefix, shown in Figure 7-2, can be read as follows: Under any circumstances, we always infer $t \bullet E \xrightarrow{t} E$. That is, an expression, with a "condition type_1 or type_2 interaction" prefixed to it, will use this "condition type_1 or type_2 interaction" to accomplish the transition.

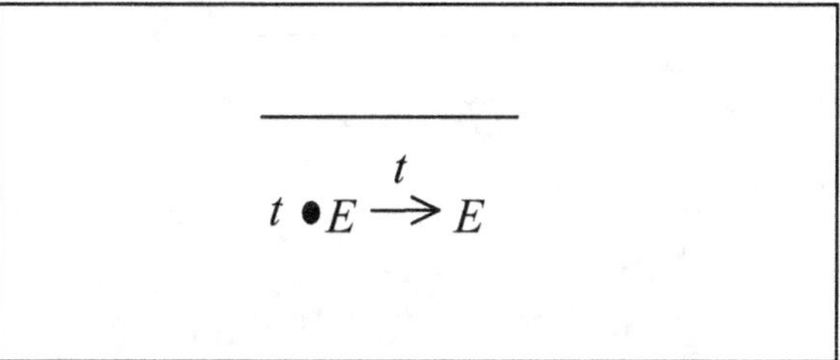

Figure 7-2 Rule of Prefix

7-3 Rule of Summation

The rule for Summation, shown in Figure 7-3, can be read as follows: If any one summand E_j of the sum $\sum_{i \in I} E_i$ has a condition type_1 or type_2 interaction, then the whole sum also has that condition type_1 or type_2 interaction.

$$\frac{E_j \xrightarrow{t} E'_j}{\sum_{i \in I} E_i \xrightarrow{t} E'_j} \quad (j \in I)$$

Figure 7-3 Rule of Summation

7-4 Rule of Parallel Composition

There are two transition rules for parallel composition. Rule Parallel$_1$, as shown in Figure 7-4, indicates that from $E \xrightarrow{t} E'$ we shall infer $E \| F \xrightarrow{t} E' \| F$.

$$\frac{E \xrightarrow{t} E'}{E \parallel F \xrightarrow{t} E' \parallel F}$$

Figure 7-4 Rule Parallel_1

Rule Parallel_2, as shown in Figure 7-5, indicates that from $F \xrightarrow{t} F'$ we shall infer $E \| F \xrightarrow{t} E \| F'$.

$$\frac{F \xrightarrow{t} F'}{E \parallel F \xrightarrow{t} E \parallel F'}$$

Figure 7-5 Rule Parallel_2

7-5 Rule of Recursion

The rule for Recursion, shown in Figure 7-6, can be read as follows: This says that any condition type_1 or type_2 interaction which may be inferred for the **fix** expression 'unwound' once (by substituting itself for its bound variable) may be inferred for the **fix** expression itself.

$$\frac{\mathbf{fix}(X{=}z\{\mathbf{fix}(X{=}z)\,/X\}) \xrightarrow{t} E'}{\mathbf{fix}(X{=}z) \xrightarrow{t} E'}$$

Figure 7-6 Rule of Recursion

7-6 Rule of Constants

The rule for Constants, shown in Figure 7-7, can be read as follows: the rule of Constants asserts that each Constant has the same transitions as its defining expression.

$$\frac{P \xrightarrow{t} P'}{A \xrightarrow{t} P'} \quad (A \overset{\text{def}}{=} P)$$

Figure 7-7 Rule of Constants

PART III: CASES STUDY

Chapter 8: Channel-Based Multi-Queue SBC Process of the Restaurant

In this chapter, we use the channel-based multi-queue SBC process algebra to define the *Restaurant* as shown in Figure 8-1.

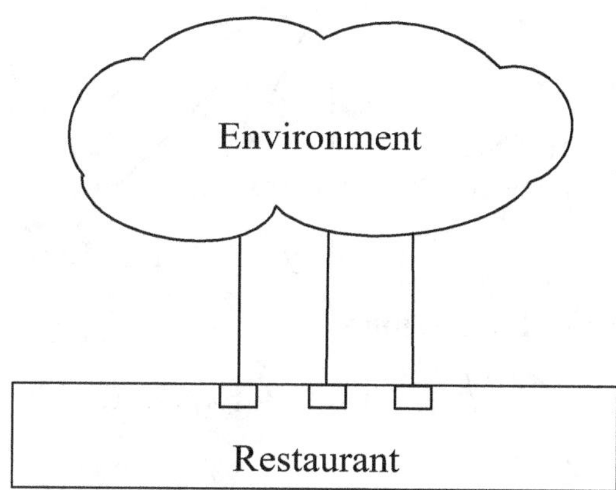

Figure 8-1 Systems Definition of the *Restaurant*

8-1 BNF Tree of the Restaurant

The channel-based multi-queue SBC process of the *Restaurant*, A_{501}, is defined as "$\mathbf{fix}(X_{501}=n_{501}\bullet(z_{502}+z_{503}+z_{504})\bullet(z_{505}+z_{506}+z_{507})\bullet n_{508}\bullet X_{501})||\mathbf{fix}(X_{502}=n_{509}\bullet X_{502})$".

We draw the channel-based multi-queue SBC process algebra Backus-Naur Form tree of A_{501} as shown in Figure 8-2.

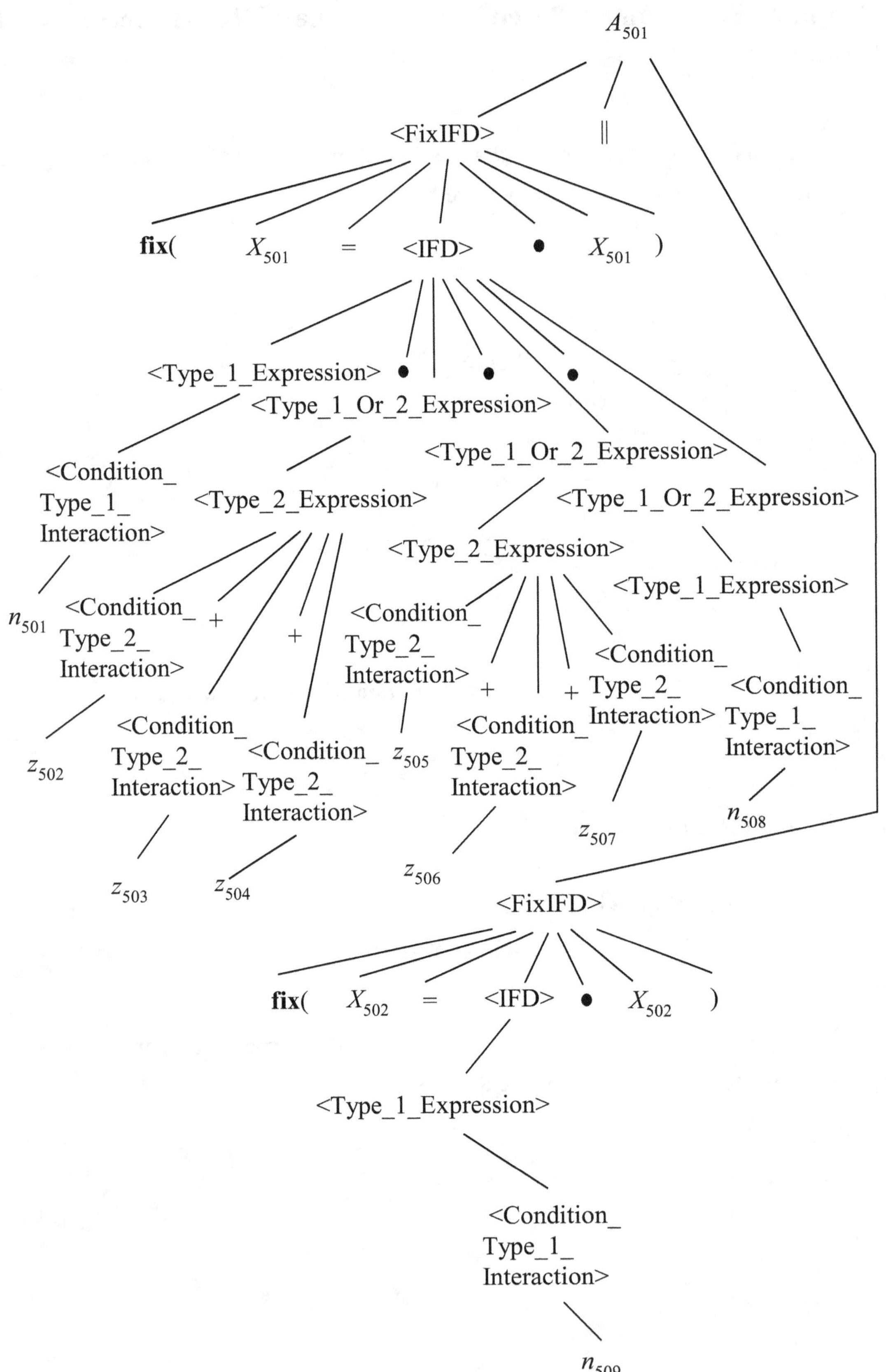

Figure 8-2 Backus-Naur Form Tree of the *Restaurant*'s Channel-Based Multi-Queue SBC Process

There are two IFDs in the channel-based multi-queue SBC process of the *Restaurant*. The first IFD is shown in Figure 8-3.

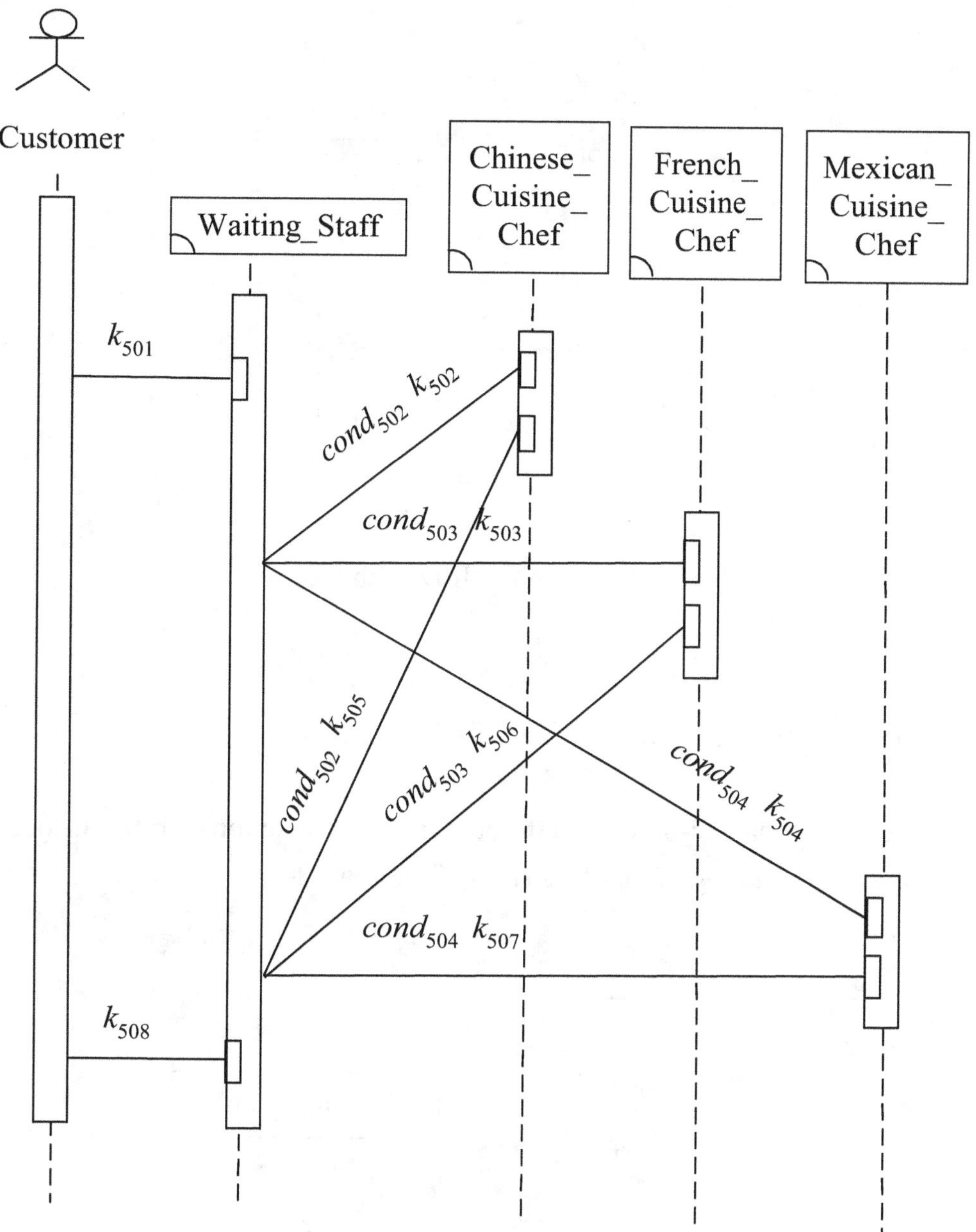

Figure 8-3 First IFD of the *Restaurant*

The second IFD of the channel-based multi-queue SBC process of the *Restaurant* is shown in Figure 8-4.

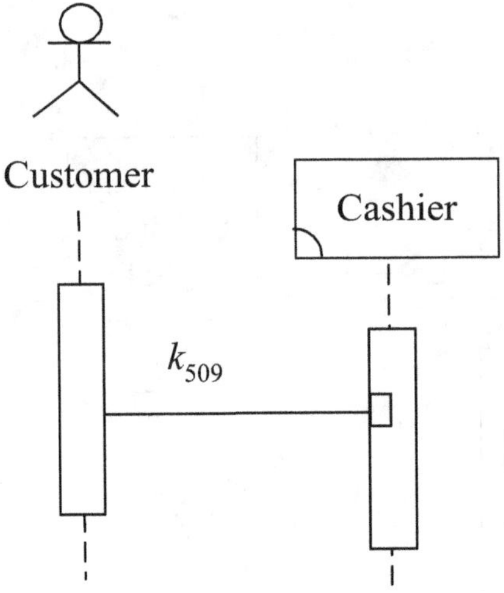

Figure 8-4 Second IFD of the *Restaurant*

8-2 Prefixes of the Restaurant

The n_{501} (channel-based condition type_1 interaction) prefix is defined as "<*Customer*, k_{501}, *Waiting_Staff*>", as shown in Figure 8-5.

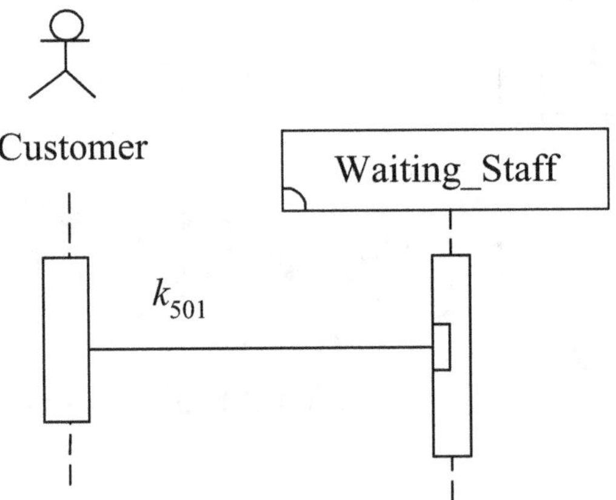

Figure 8-5 Prefix of n_{501}

The z_{502} (channel-based condition type_2 interaction) prefix is defined as "**if** Order = "Chinese" **then** <*Waiting_Staff*, k_{502}, *Chinese_Cuisine_Chef*>", as shown in Figure 8-6.

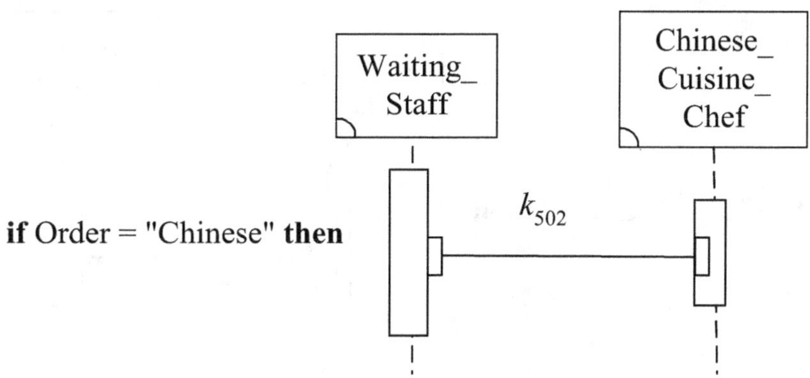

Figure 8-6 Prefix of z_{502}

The z_{503} (channel-based condition type_2 interaction) prefix is defined as "**if** Order = "French" **then** <*Waiting_Staff*, k_{503}, *French_Cuisine_Chef*>" as shown in Figure 8-7.

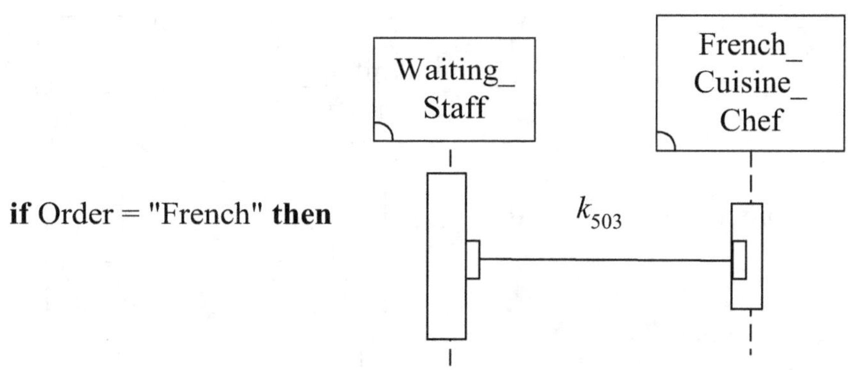

Figure 8-7 Prefix of z_{503}

The z_{504} (channel-based condition type_2 interaction) prefix is defined as "**if** Order = "Mexican" **then** <*Waiting_Staff*, k_{504}, *Mexican_Cuisine_Chef*>" as shown in Figure 8-8.

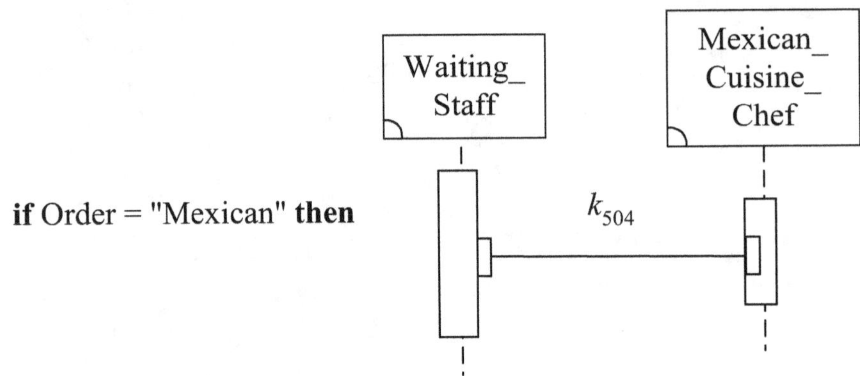

Figure 8-8 Prefix of z_{504}

The z_{505} (channel-based condition type_2 interaction) prefix is defined as "**if** Order = "Chinese" **then** <*Waiting_Staff*, k_{505}, *Chinese_Cuisine_Chef*>" as shown in Figure 8-9.

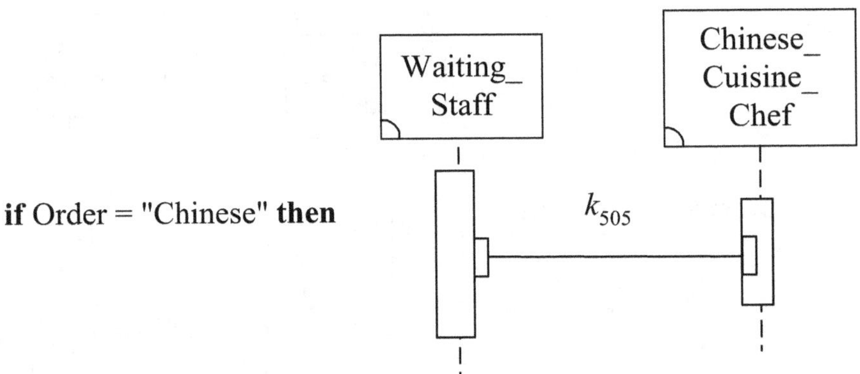

Figure 8-9 Prefix of z_{505}

The z_{506} (channel-based condition type_2 interaction) prefix is defined as "**if** Order = "French" **then** <*Waiting_Staff*, k_{506}, *French_Cuisine_Chef*>" as shown in Figure 8-10.

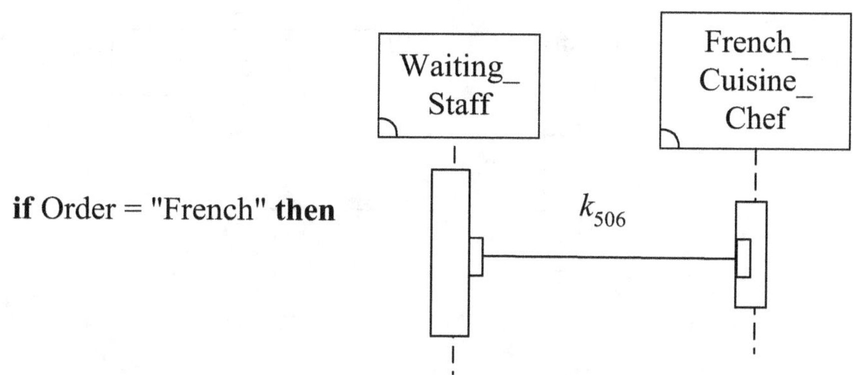

Figure 8-10 Prefix of z_{506}

The z_{507} (channel-based condition type_2 interaction) prefix is defined as "**if** Order = "Mexican" **then** <*Waiting_Staff*, k_{507}, *Mexican_Cuisine_Chef*>" as shown in Figure 8-11.

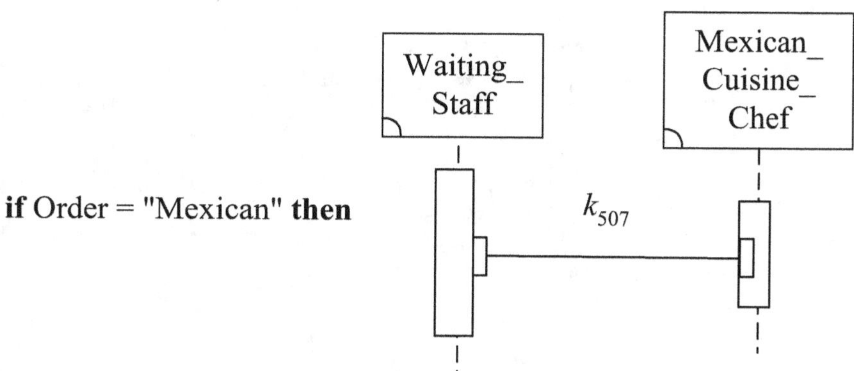

Figure 8-11 Prefix of z_{507}

The n_{508} (channel-based condition type_1 interaction) prefix is defined as "<*Customer*, k_{508}, *Waiting_Staff*>", as shown in Figure 8-12.

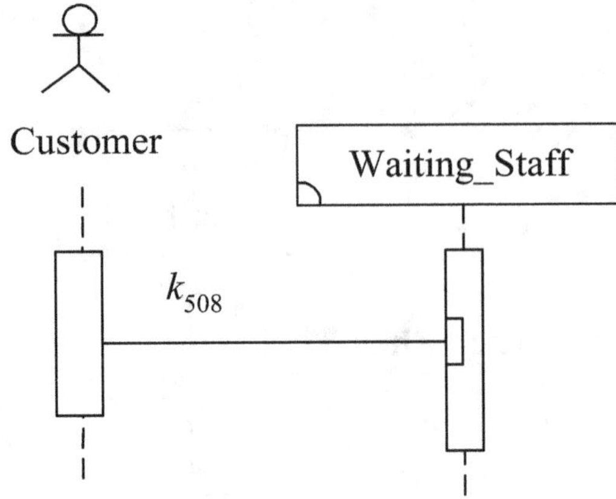

Figure 8-12 Prefix of n_{508}

The n_{509} (channel-based condition type_1 interaction) prefix is defined as "<*Customer*, k_{509}, *Cashier*>", as shown in Figure 8-13.

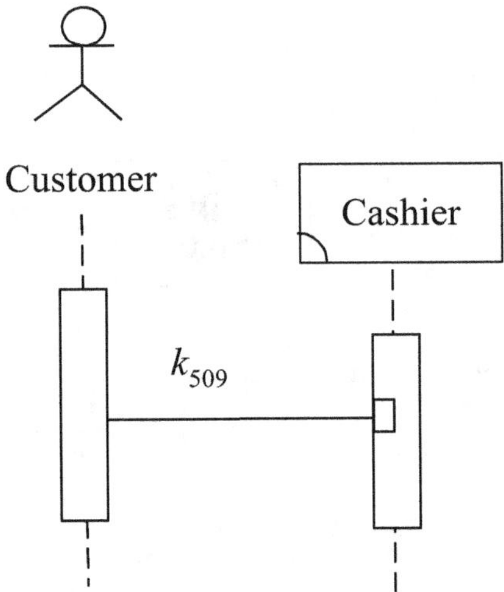

Figure 8-13 Prefix of n_{509}

Figure 8-14 shows all channel formulas of the channel-based multi-queue SBC process of the *Restaurant*.

Entity name	Channel Formula
k_{501}	Take_Order_Call(In Customer_Request)
k_{502}	Cook_Chinese_Food_Call
k_{503}	Cook_French_Food_Call
k_{504}	Cook_Mexican_Food_Call
k_{505}	Cook_Chinese_Food_Return(Out Meal_1)
k_{506}	Cook_French_Food_Return(Out Meal_2)
k_{507}	Cook_Mexican_Food_Return(Out Meal_3)
k_{508}	Take_Order_Return(Out Meal)
k_{509}	Pay_Bills

Figure 8-14 Channel Formulas of the *Restaurant*'s Process

Figure 8-15 shows the primitive data type specification of the *Order* variable, the *Customer_Request* input parameter, and the *Meal, Meal_1, Meal_2, Meal_3* output parameters.

Parameter	Data Type	Instances
Order	Enumerated	Chinese, French, Mexican
Customer_ Request	Enumerated	Chinese, French, Mexican
Meal	Enumerated food	Chinese food, French food, Mexican food
Meal_1	Food	Chinese food
Meal_2	Food	French food
Meal_3	Food	Mexican food

Figure 8-15 Primitive Data Type Specification

8-3 Process of the Restaurant

The following transition graph shows, in Figure 8-16, the semantics of P_{501}'s channel-based multi-queue SBC process.

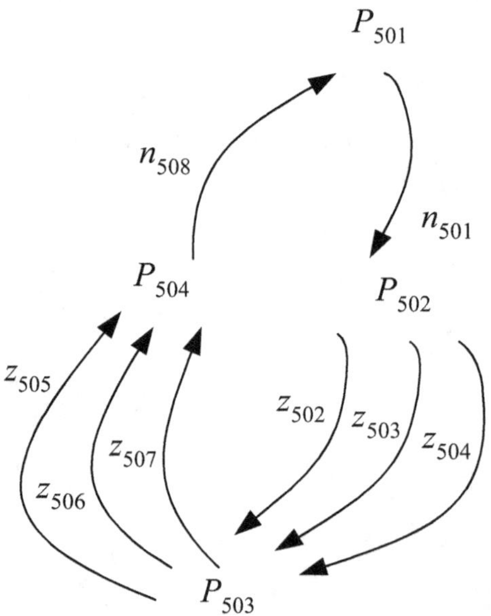

Figure 8-16 Transition graph of the P_{501} Process

In the transition graph of the P_{501}'s channel-based multi-queue SBC process, processes P_{501}, P_{502}, P_{503} and P_{504} are defined as in Figure 8-17.

$$P_{501} \overset{\text{def}}{=\!=} n_{501} \bullet P_{502}$$
$$P_{502} \overset{\text{def}}{=\!=} z_{502} \bullet P_{503} + z_{503} \bullet P_{503} + z_{504} \bullet P_{503}$$
$$P_{503} \overset{\text{def}}{=\!=} z_{505} \bullet P_{504} + z_{506} \bullet P_{504} + z_{507} \bullet P_{504}$$
$$P_{504} \overset{\text{def}}{=\!=} n_{508} \bullet P_{501}$$

Figure 8-17 Definition of Processes P_{501}, P_{502}, P_{503}, and P_{504}

The following transition graph shows, in Figure 8-18, the semantics of P_{505}'s channel-based multi-queue SBC process.

Figure 8-18 Transition graph of the P_{505} Process

In the transition graph of the P_{505}'s channel-based multi-queue SBC process, process P_{505} is defined as in Figure 8-19.

$$P_{505} \overset{\text{def}}{=\!=} n_{509} \bullet P_{505}$$

Figure 8-19 Definition of Process P_{505}

The channel-based multi-queue SBC process of the *Restaurant*, A_{501}, is defined as "$P_{501}||P_{505}$".

Chapter 9: Channel-Based Multi-Queue SBC Process of the Car

In this chapter, we use the channel-based multi-queue SBC process algebra to define the *Car* as shown in Figure 9-1.

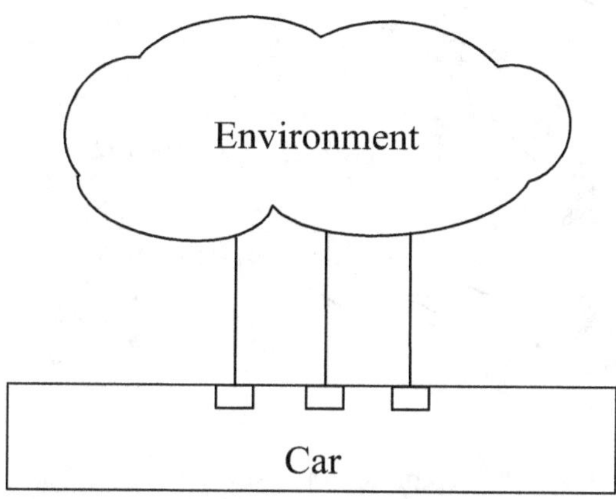

Figure 9-1 Systems Definition of the *Car*

9-1 BNF Tree of the Car

The channel-based multi-queue SBC process of the *Car*, A_{601}, is defined as "$\mathbf{fix}(X_{601}=n_{601}\bullet X_{601})||\mathbf{fix}(X_{602}= n_{602}\bullet X_{602})||\mathbf{fix}(X_{603}=n_{603}\bullet z_{604}\bullet(z_{605}+z_{606})\bullet X_{603})$".

We draw the channel-based multi-queue SBC process algebra Backus-Naur Form tree of A_{601} as shown in Figure 9-2.

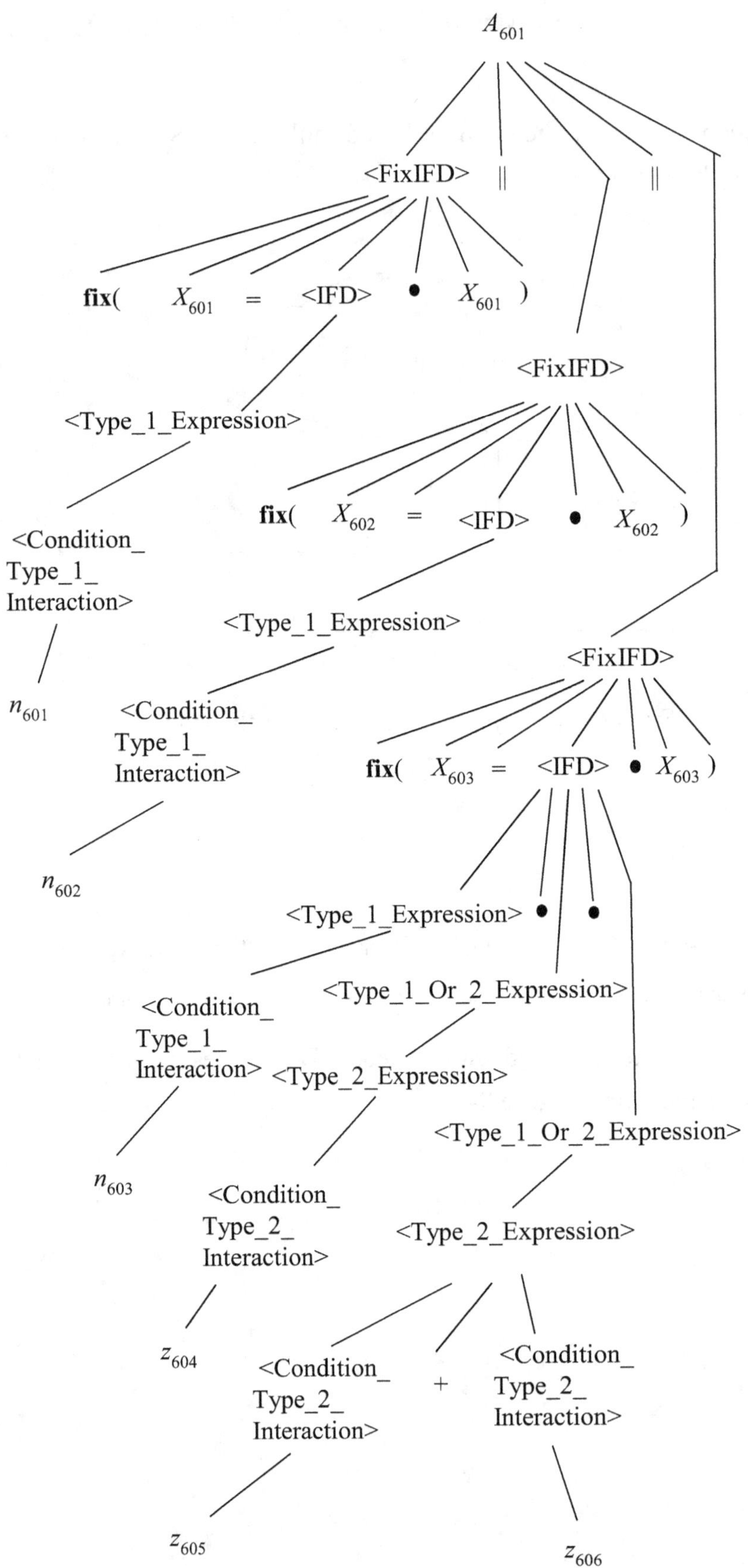

Figure 9-2 Backus-Naur Form Tree of the *Car*'s Channel-Based Multi-Queue SBC Process

There are three IFDs in the channel-based multi-queue SBC process of the *Car*. The first IFD is shown in Figure 9-3.

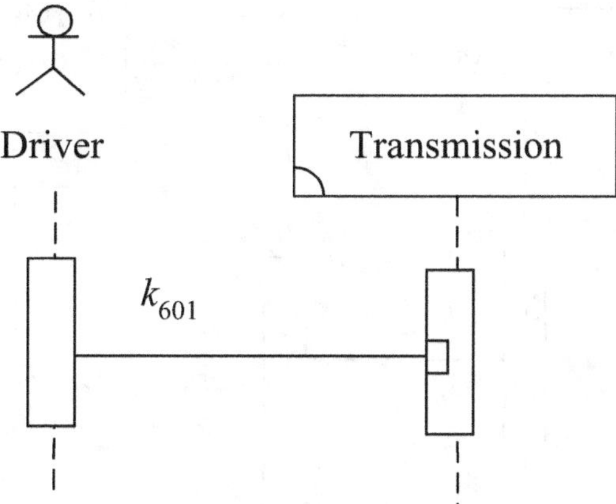

Figure 9-3 First IFD of the *Car*

The second IFD of the channel-based multi-queue SBC process of the *Car* is shown in Figure 9-4.

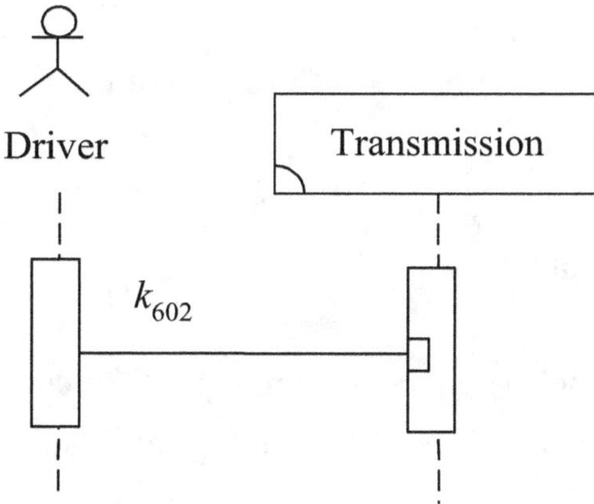

Figure 9-4 Second IFD of the *Car*

The third IFD of the channel-based multi-queue SBC process of the *Car* is shown in Figure 9-5.

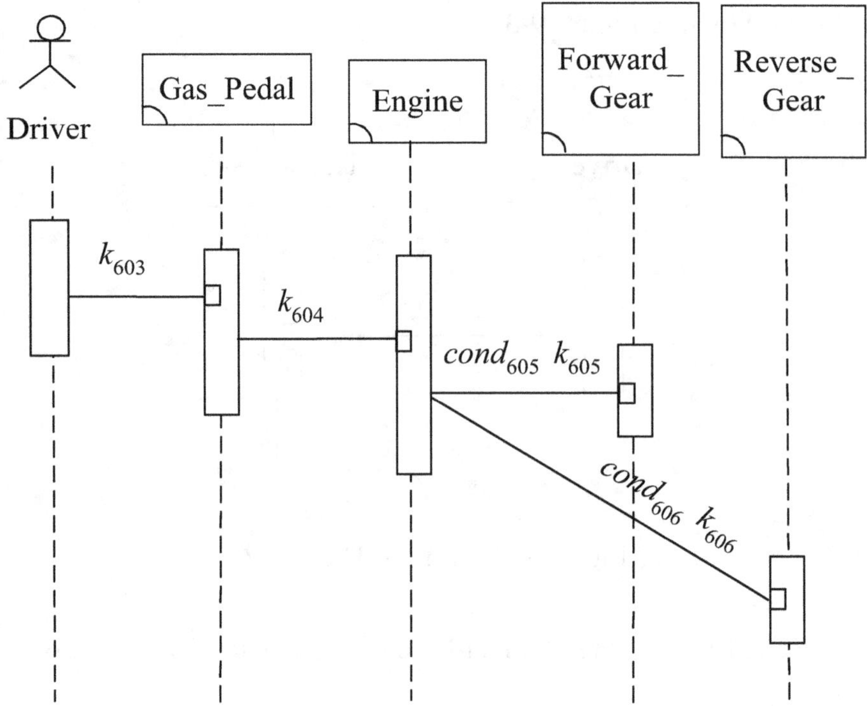

Figure 9-5 Third IFD of the *Car*

9-2 Prefixes of the Car

The n_{601} (channel-based condition type_1 interaction) prefix is defined as "<*Driver*, k_{601}, *Transmission*>", as shown in Figure 9-6.

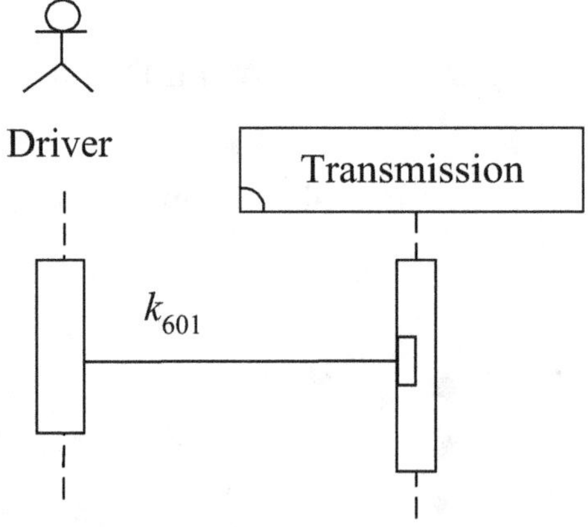

Figure 9-6 Prefix of n_{601}

The n_{602} (channel-based condition type_1 interaction) prefix is defined as "<*Driver*, k_{602}, *Transmission*>", as shown in Figure 9-7.

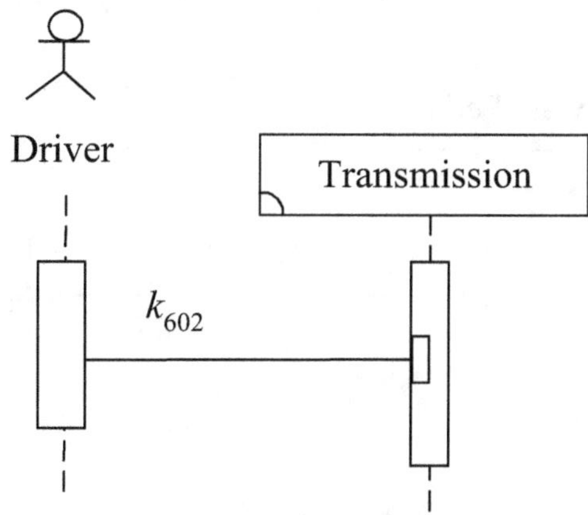

Figure 9-7 Prefix of n_{602}

The n_{603} (channel-based condition type_1 interaction) prefix is defined as "<*Driver*, k_{603}, *Gas_Pedal*>", as shown in Figure 9-8.

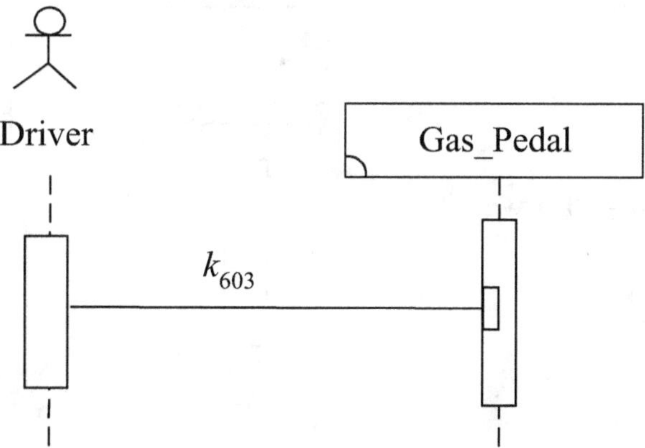

Figure 9-8 Prefix of n_{603}

The z_{604} (channel-based condition type_2 interaction) prefix is defined as "<*Gas_Pedal*, k_{604}, *Engine*>", as shown in Figure 9-9.

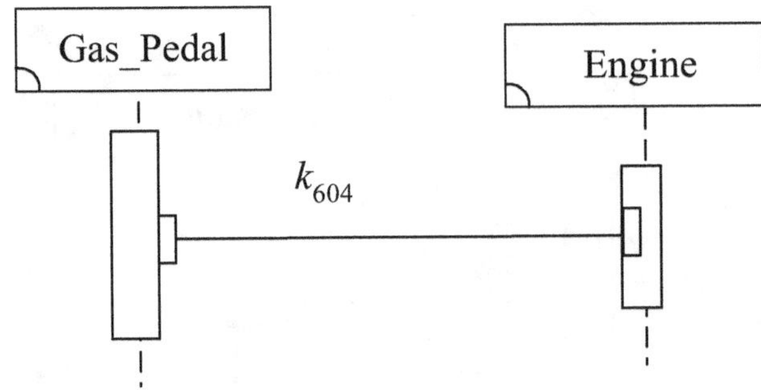

Figure 9-9 Prefix of z_{604}

The z_{605} (channel-based condition type_2 interaction) prefix is defined as "**if** Gear = "Forward" **then** <*Engine*, k_{605}, *Forward_Gear*>" as shown in Figure 9-10.

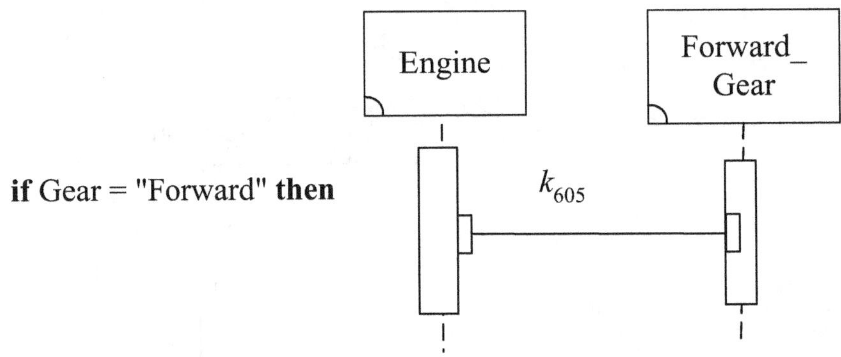

Figure 9-10 Prefix of z_{605}

The z_{606} (channel-based condition type_2 interaction) prefix is defined as "**if** Gear = "Reverse" **then** <*Engine*, k_{606}, *Reverse_Gear*>" as shown in Figure 9-11.

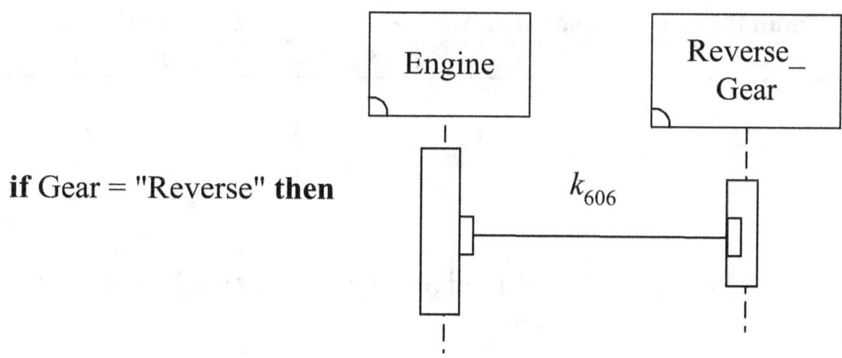

Figure 9-11 Prefix of z_{606}

Figure 9-12 shows all channel formulas of the channel-based multi-queue SBC process of the *Car*.

Entity name	Channel Formula
k_{601}	Push_Gear_Forward
k_{602}	Push_Gear_Reverse
k_{603}	Depress_Gas_Pedal
k_{604}	Fuel_Supply
k_{605}	Forward_Gear_Rotate
k_{606}	Reverse_Gear_Rotate

Figure 9-12 Channel Formulas of the *Car*'s Process

Figure 9-13 shows us the primitive data type specification of the *Gear* variable.

Parameter	Data Type	Instances
Gear	Enumerated	Forward, Reverse

Figure 9-13 Primitive Data Type Specification

9-3 Process of the Car

The following transition graph shows, in Figure 9-14, the semantics of P_{601}'s channel-based multi-queue SBC process.

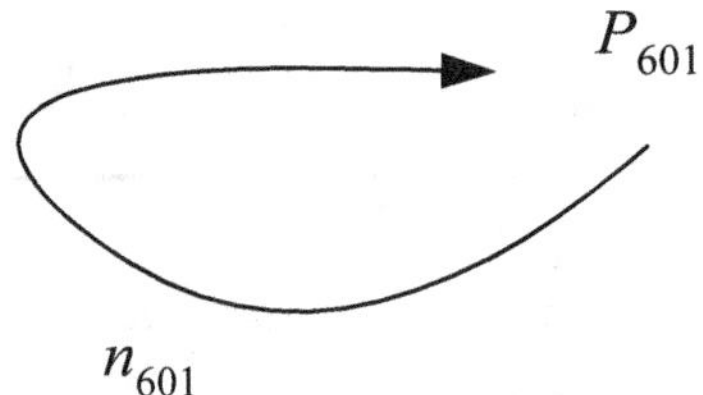

Figure 9-14 Transition graph of the P_{601} Process

In the transition graph of the P_{601}'s channel-based multi-queue SBC process, process P_{601} is defined as in Figure 9-15.

$$P_{601} \stackrel{\text{def}}{=\!=} n_{601} \bullet P_{601}$$

Figure 9-15 Definition of Process P_{601}

The following transition graph shows, in Figure 9-16, the semantics of P_{602}'s channel-based multi-queue SBC process.

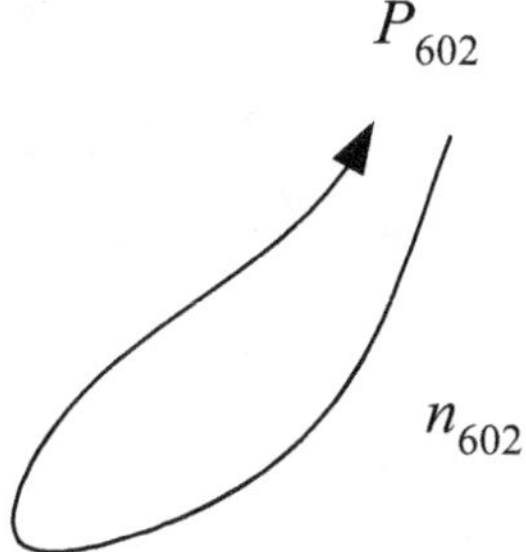

Figure 9-16 Transition graph of the P_{602} Process

In the transition graph of the P_{602}'s channel-based multi-queue SBC process, process P_{602} is defined as in Figure 9-17.

$$P_{602} \stackrel{\text{def}}{=\!=} n_{602} \bullet P_{602}$$

Figure 9-17 Definition of Process P_{602}

The following transition graph shows, in Figure 9-18, the semantics of P_{603}'s channel-based multi-queue SBC process.

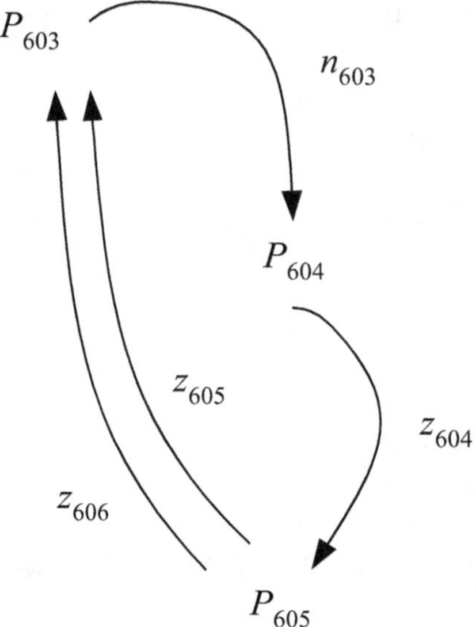

Figure 9-18 Transition graph of the P_{603} Process

In the transition graph of the P_{603}'s channel-based multi-queue SBC process, processes P_{603}, P_{604} and P_{605} are defined as in Figure 9-19.

$$P_{603} \overset{\text{def}}{=\!=} n_{603} \bullet P_{604}$$
$$P_{604} \overset{\text{def}}{=\!=} z_{604} \bullet P_{605}$$
$$P_{605} \overset{\text{def}}{=\!=} z_{605} \bullet P_{603} + z_{606} \bullet P_{603}$$

Figure 9-19 Definition of Processes P_{603}, P_{604}, and P_{605}

The channel-based multi-queue SBC process of the *Car*, A_{601}, is defined as "$P_{601}||P_{602}||P_{603}$".

APPENDIX A: Language Constructs of Channel-Based Multi-Queue SBC Process Algebra

(1) <System> ::= <FixIFD> {“|| ” <FixIFD>}

(2) <FixIFD> ::= “fix(” <Process_Variable>“=”<IFD>
“●” <Process_Variable> “)”

(3) <IFD> ::= <Type_1_Expression> {“●” <Type_1_Or_2_Expression>}

(4) <Type_1_Or_2_Expression> ::= <Type_1_Expression>

| <Type_2_Expression>

(5) <Type_1_Expression> ::= <Condition_Type_1_Interaction>
{“+” <Condition_Type_1_Interaction>}

(6) <Type_2_Expression> ::= <Condition_Type_2_Interaction>
{“+” <Condition_Type_2_Interaction>}

APPENDIX B: Transitional Semantics of Channel-Based Multi-Queue SBC Process Algebra

Prefix

$$\frac{}{t \bullet E \xrightarrow{t} E}$$

Sum_j

$$\frac{E_j \xrightarrow{t} E'_j}{\sum_{i \in I} E_i \xrightarrow{t} E'_j} (j \in I)$$

Parallel_1

$$\frac{E \xrightarrow{t} E'}{E \parallel F \xrightarrow{t} E' \parallel F}$$

Parallel_2

$$\frac{F \xrightarrow{t} F'}{E \parallel F \xrightarrow{t} E \parallel F'}$$

Recursion

$$\frac{\mathbf{fix}(X{=}z\{\mathbf{fix}(X{=}z)/X\}) \xrightarrow{t} E'}{\mathbf{fix}(X{=}z) \xrightarrow{t} E'}$$

Constant

$$\frac{P \xrightarrow{t} P'}{A \xrightarrow{t} P'} \quad (A \overset{\text{def}}{=\!=} P)$$

BIBLIOGRAPHY

[Acko68] Ackoff, R., "Toward a System of Systems Concepts," *Modern Systems Research for the Behavioral Scientist: A Sourcebook*, Aldine Publishing Company, 1968.

[Berg87] Bergstra, J. A. et al., "ACPτ: A Universal Axiom System for Process Specification," *CWI Quarterly* 15, 1987, pp. 3-23.

[Burd10] Burd, S. D., *Systems Architecture*, 6th Edition, Cengage Learning, 2010.

[Chao14a] Chao, W. S., *Systems Thingking 2.0: Architectural Thinking Using the SBC Architecture Description Language*, CreateSpace Independent Publishing Platform, 2014.

[Chao14b] Chao, W. S., *General Systems Theory 2.0: General Architectural Theory Using the SBC Architecture*, CreateSpace Independent Publishing Platform, 2014.

[Chao14c] Chao, W. S., *Systems Modeling and Architecting: Structure-Behavior Coalescence for Systems Architecture*, CreateSpace Independent Publishing Platform, 2014.

[Chao15a] Chao, W. S., *Variants of Interaction Flow Diagrams: The Structure-Behavior Coalescence Approach*, CreateSpace Independent Publishing Platform, 2015.

[Chao15b] Chao, W. S., *A Process Algebra For Systems Architecture: The Structure-Behavior Coalescence Approach*, CreateSpace Independent Publishing Platform, 2015.

[Chao15c] Chao, W. S., *An Observation Congruence Model For Systems Architecture: The Structure-Behavior Coalescence Approach*, CreateSpace Independent

Publishing Platform, 2015.

[Chao15d] Chao, W. S., *Variants of SBC Process Algebra: The Structure-Behavior Coalescence Approach*, CreateSpace Independent Publishing Platform, 2015.

[Chao15e] Chao, W. S., *Single-Queue SBC Process Algebra For Systems Architecture: The Structure-Behavior Coalescence Approach*, CreateSpace Independent Publishing Platform, 2015.

[Chao15f] Chao, W. S., *Multi-Queue SBC Process Algebra For Systems Architecture: The Structure-Behavior Coalescence Approach*, CreateSpace Independent Publishing Platform, 2015.

[Chao15g] Chao, W. S., *Single-Queue SBC Observation Congruence Model For Systems Architecture: The Structure-Behavior Coalescence Approach*, CreateSpace Independent Publishing Platform, 2015.

[Chao15h] Chao, W. S., *Multi-Queue SBC Observation Congruence Model For Systems Architecture: The Structure-Behavior Coalescence Approach*, CreateSpace Independent Publishing Platform, 2015.

[Chao16a] Chao, W. S., *System: Contemporary Concept, Definition, and Language*, CreateSpace Independent Publishing Platform, 2016.

[Chao16b] Chao, W. S., *Systems Architecture of Electronic Toll Collection Cloud Applications and Services IoT System*, CreateSpace Independent Publishing Platform, 2016.

[Chao16c] Chao, W. S., *Systems Architecture of Smart Healthcare Cloud Applications and Services IoT System*, CreateSpace Independent Publishing Platform, 2016.

[Chao16d] Chao, W. S., *Systems Architecture of Smart Healthcare Cloud Applications and Services IoT System*, CreateSpace Independent Publishing

Platform, 2016.

[Chao16e] Chao, W. S., *Systems Architecture of Ridesharing Sharing Economy Cloud Applications and Services IoT System*, CreateSpace Independent Publishing Platform, 2016.

[Chao16f] Chao, W. S., *Systems Architecture of Handy Helper Sharing Economy Cloud Applications and Services IoT System*, CreateSpace Independent Publishing Platform, 2016.

[Chao16g] Chao, W. S., *Systems Architecture of Vacation Rental Cleaning Sharing Economy Cloud Applications and Services IoT System*, CreateSpace Independent Publishing Platform, 2016.

[Chec99] Checkland, P., *Systems Thinking, Systems Practice: Includes a 30-Year Retrospective*, 1st Edition, Wiley, 1999.

[Craw15] Crawley, P. et al., *System Architecture: Strategy and Product Development for Complex Systems*, Prentice Hall, 2015.

[Dam06] Dam, S., *DoD Architecture Framework: A Guide to Applying System Engineering to Develop Integrated Executable Architectures*, BookSurge Publishing, 2006.

[Date03] Date, C. J., *An Introduction to Database Systems*, 8th Edition, Addison Wesley, 2003.

[Elma10] Elmasri, R., *Fundamentals of Database Systems*, 6th Edition, Addison Wesley, 2010.

[Frie11] Friedenthal, S., et al., *A Practical Guide to SysML: The Systems Modeling Language*, Morgan Kaufmann, 2nd Edition, 2011.

[Ghar11] Gharajedaghi, J., *Systems Thinking: Managing Chaos and Complexity: A Platform for Designing Business Architecture*, Morgan Kaufmann, 2011.

[Hoar85] Hoare, C. A. R., *Communicating Sequential Processes*, Prentice-Hall, 1985.

[Maie09] Maier, M. W., *The Art of Systems Architecting*, 3rd Edition, CRC Press, 2009.

[Mead08] Meadows, D. H., *Thinking in Systems: A Primer*, Chelsea Green Publishing, 2008.

[Miln89] Milner, R., *Communication and Concurrency*, Prentice-Hall, 1989.

[Miln99] Milner, R., *Communicating and Mobile Systems: the π-Calculus*, 1st Edition, Cambridge University Press, 1999.

[O'Rou03] O'Rourke, C. et al, *Enterprise Architecture Using the Zachman Framework*, 1st Edition, Course Technology, 2003.

[Putm00] Putman, J. R. et al., *Architecting with RM-ODP*, Prentice-Hall, 2000.

[Rayn09] Raynard, B., *TOGAF The Open Group Architecture Framework 100 Success Secrets*, Emereo Pty Ltd, 2009.

[Roza11] Rozanski, N. et al., *Software Systems Architecture: Working With Stakeholders Using Viewpoints and Perspectives*, 2nd Edition, Addison-Wesley Professional, 2011.

[Scho10] Scholl, C., *Functional Decomposition with Applications to FPGA Synthesis*, Springer, 2010.

[Toga08] The Open Group, *TOGAF Version 9 - A Manual (TOGAF Series)*, 9th Edition, Van Haren Publishing, 2008.

INDEX

P

R

S

T

www.ingramcontent.com/pod-product-compliance
Lightning Source LLC
Chambersburg PA
CBHW081201180526
45170CB00006B/2176

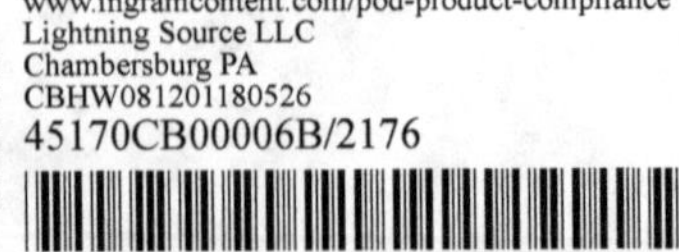

* 9 7 8 1 5 4 6 4 6 5 2 5 6 *